BRAVE NEW MIND

Living in a Future-Science World

Elliott Maynard, Ph.D.

Arcos Cielos Publishers
P.O. Box 20069
Sedona, Arizona 86341
www.arcoscielos.com

Brave New Mind
Living in a Future-Science World
by Elliott Maynard, Ph.D.

Published by
Arcos Cielos Publishers
P.O. Box 20069
Sedona, Arizona 86341, USA
www.arcoscielos.com

Printed in the United States of America.
All rights reserved.
No part of this book shall be reproduced in any form
except for the inclusion of brief quotations
covered under "Fair Use of Copyright Law"
without written permission from the author or publisher.

Copyright© 2014 by Elliott Maynard

ISBN #978-0-9960780-0-9

Cover Photo: Willyam Bradberry

Layout and Design: Karen Reider

Author Photos: Michael Helms

Artwork, *Human from the Future:* Abe Kreworuka

Other Books by Elliott Maynard

Transforming the Global Biosphere
Twelve Futuristic Strategies

Life Management 3000
Success & Survival in the Third Millennium

Future-Science Art
A Unique Paradigm for Creating
"Living Artwork"

To purchase or learn more about
these books, please go to:
www.arcoscielos.com

Dedication

To my wife Alisa, for her loving support,
editing brilliance and dogged determination.
And to the tourchbearers of today
and the generations yet to come.
May they lead us to ever brighter pathways
into a sustainable future – for humans
and for Planet Earth.

Acknowledgments

Team Avanti:
Steve Brown, Fulfillment
Porsche Gallucci, Creative Development
Ingrid Hardy, Personal Assistant
Alisa F. Maynard, Executive Producer, Agent, Management
Light Maynard, Spirit/Medicine Cat
Pavel Mikoloski, Public Relations

• • •

Dr. Eben Alexander, Neurosurgeon/Author, "Proof of Heaven"
Allen H. Arrow, Esq., Legal Counsel & Arbiter of Great Good Will
Lt. Col. Thomas F. Bearden (U.S. Army, Ret.), Scalar Technology, Zero-Point Energy Pioneer
Bonnie Bennett, D.O., Homeopath/Osteopath/Wholistic Medical Educator
Boyd Bushman, Lockheed Martin Senior Scientist
Kim Carlsberg, UFO Contactee/Author, "The Art of Close Encounters"
President Jimmy Carter & Rosalyn Carter
Jesai Chantler, Astrologer, After-Death Communicator
Cybela Clare, Filmmaker/Prism Films, "Bird's Eye View" and "Animal Magnetism"
Portlin Cochise, Author, Neighbor Extraordinaire
Kenneth Cox, Ph.D., former NASA Apollo Engineer, Leader: Aerospace Technology Working Group (ATWG)
Carson Davidson, Author/Editor/Filmmaker, Carson Davidson Productions
Robert O. Dean, Ufologist/Alien Contactee (U.S. Army, Ret.)
Valmai Howe Elkins, Author, Feng Shui Master, Neighbor Extraordinaire
Paola Farina, Parapsychology Student/Researcher
Jacque Fresco, Futuristic Architect/Engineer – The Venus Project
Albert Clayton Gaulden, Master Astrologer, Author/Founder – The Sedona Intensive
Russ George, Global Ocean Scientist
Geoffrey Gordon, Musician/Rhythms From Beyond the Veil (Breath of Rama)
Charles J. Hall, Nuclear Physicist, Alien Contact Specialist
Mary Hardy, Earth Geomancy
Michael Helms, Michael Helms Photography
Paul Horn, TM Meditation Teacher, "Father of New Age Music," Flutist
Susan Hutchins, Medical Intuitive, Dark-Field Live Blood Cell Analysis

Grand Master Darryl Khalid, Iron Shirt Chi Gong, Energy Specialist
Nawang Khechog, Tibetan Flutist, Author, "Awakening Kindness"
Abe Kreworuka, Maine Folk Artist, Best Friend
Karen Kreworuka, Editor, Maine Blueberry Crumble Diva
Peter F. Langrock, Vermont Attorney, Author, "Addison County Justice"
LionFire Leonard, Artist, Shaman, Musician
Mary Leonard, Cloud-Woman, Healer
Gayle Mack, Naturopath
Kim McDermott, Yoga Teacher, Media Professional
Roxanne Meadows, Futuristic Architect, The Venus Project
Andrew Michrowski, Founder/President – Planetary Association for Clean Energy
Robert Miles, Film Producer of ET/UFO Movie "Fastwalkers"
Judith Moore, Professional Channeler & Psychic
Ann Mortifee, Musician/Composer, Author, "In Love With the Mystery"
Robert Muller, Ph.D., Former U.N. Assistant Secretary General
Sal Rachelle, Professional Channeler, Spititual Guide
Karen V. Reider, Artist, Editor, Book Design Specialist
Terra Sonora, Professional Channeler, Spiritual Guide
Gerard Ungerman, Filmmaker, The Respectful Revolution
Tom Valone, Ph.D., Founder/President, Integrity Research Institute
Stacey Weare, Filmmaker, The Respectful Revolution
Jerry Wills, Healer, Medical Intuitive, Musician
Kathy Wills, Healer, Herbalist
Edward Winchester, Founder, The Pentagon Meditation Club & Peacemakers' Institute
Venerable Tenzin Yignyen, Professor, Tibetan Buddhism; Tibetan Sand Mandala Artist

In Memoriam:
Paul Horn, Geoffrey Gordon,
Sharon Midori Maynard,
Robert Muller, Dr. Brian O'Leary,
Allan Silberhartz

Table of Contents

Foreword viii

Introduction x

Part I 1
   **Working with the Quantum Field, Linear vs Quantum Thinking
   from a Future-Science Technology Perspective**
Collective Consciousness: A Powerful Resource for Igniting Positive Change 3
Internet-Based Torsionic Meditation: A Mechanism for Shifting
   the Global Consciousness 5
A New Format for Future-Science Paradigms 7
Old Paradigms: Examples of the Industrial-Age Mindset 7
New Future-Science Paradigms as Triggers for Consciousness Transformation
   and Human Re-Evolution 8
The Breakdown of Conventional Scientific Thinking: Re-Inventing Science
   to Fit the Times 9
"Brave New Mind," A New Operating System for the Future 10
New Pathways for Human Enlightenment and Biospherical Restoration 11

Part II 14
   The Core Paradigms of Future-Science Technology
Master Key 1 - Supersensonics: A Tool for Evolving Human Potential 14
Nuclear Evolution and the Antakarana System 15
The Art and Science of Dowsing: A Key Aspect of Supersensonics 16
Dowsing: An Ancient Science for Amplifying the Intuitive Senses
   and Interfacing with the Quantum Information Field 18
Final Thoughts on Supersensonics and Dowsing 22
Master Key 2 - Human and Planetary Eco-Evolution 22
The Gaia Hypothesis, Global Ecology and Future-Science Technology 23
Environmental Generational Amnesia 23
The Healing Effects of Forests and the Natural World 25
Man's Oldest Ancestor More than Twice as Old as Previous Estimates 26
Greenland Ice Sheet Melting 27
Earth's Glaciers Are Out of Balance 28
Global Weather Engineering: The History of Rainmaking...28
What in the World Are They Spraying? The Chemtrail Conspiracy..........................30
Wind Turbines Can Modify the Weather, but Do Long-Term Benefits Outweigh
   the Environmental Consequences? 31
Negative Impacts of Wind Turbines and the "Wind Turbine Syndrome" 32
The Oldest Trees on Earth 34
Scientists Revive a Flowering Plant after 30,000 Years 34
Extreme Life-Forms Redefine Our Definition of "Life" and Offer
   New Implications for Astrobiologists 35
Why Whales are a Key to Sustainable Ocean Fisheries 38
**Master Key 3 - Future-Science Education: Creating New
   Learning Systems for Coping with the Present and Future Realities** 40
The University of the Future Concept 42

Communications Fluency and Conscious Evolution 43
Cyberlearning 44
The Global Internet and "Webucation" 46
Virtual Universities Gain New Ground 47
Creating a Synergistic Future-Science Learning Community 48
Subtle-Energy Management: Key to Future Education 49
Rewiring Your Brain to Learn a Language in 10 Days 50
Conclusions and Final Thoughts 51

Master Key 4 - Future-Science Art: A Quantum Technology
   for Creating "Living Artworks" 52
The Historical Background of Future-Science Art 53
Techniques for Creating Future-Science Sculptures 54
Components Used in Creating Future-Science Sculptures 55
Consciousness Techniques for Creating Future-Science Artworks 56
Manifesting Future-Science Artworks: The Co-Creative Process 57
Additional Aspects of the Creative Process 59
Final Thoughts on Future-Science Art 60

Master Key 5 - Future-Science Music 61
How Music Tones the Brain and Improves Learning 64
Musical Duets Lock Brains and Rhythms 65
The Amazing Powers of Music Revealed 66
Visualization Plus Vocalization Equals Manifestation 67
Consciousness Attunement and DNA Activation via the Sacred Solfeggio Scale 68

Master Key 6 - Earth Energies 70
Magnetic Therapy for Healing and Sleep Enhancement 71
New Future-Science Technology Applications for Earth Energies 74

Master Key 7 - Psychic Cleanliness and Conscious Evolution 74
Learning to Shift Frequencies 74
Cleaning Our Energy Fields and Shifting into an Interdimensional
   Quantum Consciousness Framework 75
Psychic Cleanliness: A Consciousness Evolution 76
Overcoming Tribal Beliefs Embedded in the Subconscious Mind 78

Master Key 8 - Tesla Technology 79
Tesla Technology: A Quantum Field Tool for Healing 80
The Multiwave Oscillator: A Healing System Based on Tesla Technology 81
The HAARP Project: An Example of Tesla Technology in Action 82

Master Key 9 - ELF Electronics 85
Schumann Resonance: The Pulse of Planet Earth 85
ELF Waves and Human Brain Waves: How They Interact 87

Master Key 10 - Radionics: A Quantum Technology for
   Human and Planetary Healing 89
Radionics: A Quantum-Field Healing Technology 89
Radionic Aquaculture: A Futuristic Paradigm for Intensive Food Production 91
Some Proposed Futuristic Application for Radionic Technologies 92
A Modest Proposal: A Future-Science Approach to Planetary Management 93

Master Key 11 - New Futuristic Technologies 94
Non-Conventional Technologies as Change Agents for Human
   and the Global Biosphere 94
A Car that Needs Refueling Every 100 Years 96
3D Printing: An Amazing New Technology Already Up and Running 97
A New 3D Printer that Uses Embryonic Stem Cells..100
Building a House in 20 Hours with 3D Printing..100
The Amaze Project Brings 3D Printing into the Space Age 101
3D Printed Buildings for the Moon 102
The Urbee Hybrid: The First 3D Printed Car 103
A Promising Technology for Transforming Air Pollution into Baking Soda.............104
Bioplastics: An Eco-Sensible Strategy for Reducing Plastic Waste 105
Converting Waste Plastics into Diesel Fuel 106
Liquid Air Technology: A New Way to Store Intermittent Solar and Wind Energy 108
Multiuse Titanium Dioxide: New Compound that Can Generate Hydrogen
   and Produce Clean Water While Processing Wastewater 109
Thorium Nuclear Reactors: A Bridging Technology to Bypass Dangers
   of Conventional Nuclear Reactor Technology 111
Precursor Engineering as an Aspect of Future-Science Energy Technologies...........113
Master Key 12 - Interspecies Communication and Interaction 115
Animal Communicators: Creating a Bridge Between Humans
   and Other Species 115
South African Animal Communicator, Anna Breytenbach 116
The Passing of My 19-Year-Old Cat Companion 117
Master Key 13 - Subtle-Energy Management 119
When You Change the Energy You Change the Mass 119
The Hundredth Monkey Effect as an Aspect of Quantum-Field Technology............120
Financial Energy Management: Money as "Frozen Energy" 121
The Physical Plane as a Training Ground: Learning to Control Energy
   and Observe the Results 121
Managing Energy Fields and Creating Sacred Space through Visualization
   and Intent 122
Shifting Frequencies, Shifting Fields..126
Spiritual Responsibility and Guidance from the Higher Realms................................126
Master Key 14 - AI/Human Energy-Field Interfacing 127
Is the Internet Becoming A Conscious Entity? 127
Digital Prayer Wheels and Torsionic Meditation..128
The "Force-Magnifier Effect:" An Effective Tool for Social Transformation.............129
Welcome to the Future of Medicine! Star Trek Technology has Arrived! 130
Electronics as Spiritual Wave Guides 132
Master Key 15 - Future-Science Medicine..133
A Tiny Fish that Could Transform Biology and Medicine 134
The Role of Medical Intuitives in Future-Science Medicine..135
Personal Experiences in Intuitive and Modern Medicine 137
Medical Intuitive and Healer, Jerry Wills 140
Future-Science Healing Protocols Based on Intuitive Medicine 143
Curing Cancer, Radiation Poisoning and Other Medical Conditions with Cannabis:
   A Surprising Natural Medicine 144

Major Breakthrough in Stem Cell Therapy 145
Possession Therapy 146
Accepting Responsibility, Overcoming Tribal Beliefs
  and Healing Ourselves through Our Thoughts and Feelings 147
Human/Extraterrestrial Healing 148
Israel's Extraterrestrial Healing Clinic 148
Final Thoughts on Future-Science Medicine 150
Master Key 16 - Conscientious Biotechnology 150
Repelling Viruses, Reviving Mammoths, Cloning "Supertrees" 150
Amazing Test Tube Forests 153
Restoring Our Natural Ecological Heritage by Cloning Redwoods 154
Producing Human Stem Cells Without Embryos 155
Genetically Modified Salmon Awaiting FDA Approval for Human Consumption 157
Future-Science Applications of Conscientious Genetic Engineering 160
**Master Key 17 - Super-Ecology: A New Future Paradigm
  for Sustainable Living** 160
Super-Ecology: Restoring Planet Earth's Natural Resources
Restorative Aquaculture: Strategies for Marine Environmental Resource
  Management 163
Sustainable Aquafarms and "Environmental Tithing" 165
Great Barrier Reef Corals are Disappearing at an Alarming Rate 166
Super-Ecology Strategies for Coral Reef Restoration 167
Seeding Ocean Currents with Powdered Iron to Create Plankton
  Blooms and Enrich Food Chains for Ocean Fisheries 169
Final Thoughts on Super-Ecology 174
**Master Key 18 - Harmonic Attunement Technology: A New Method
  to Upgrade Human and Planetary Consciousness** 175
Harmonic Attunement Technology: A New Paradigm that Integrates "Resonance"
  and "Consciousness Entrainment" 175
Final Thoughts on Harmonic Attunement Technology 179
Master Key 19 - Superfoods and Supernutrition 180
Superfoods: What are They and Where Do They Come From? 180
The "Living Lunchbox" Concept 181
Designer Eggs: An Extension of the Living Lunchbox Concept 182
Producing Higher Quality Fish with Designer Diets 183
Biodynamic Agriculture: A Natural Way to Produce Superfoods 185
Paramagnetic Agriculture as an Aspect of Superfood Production 186
In-Vitro Culture and Bioreactors: Food Production for the Future? 188
Spirulina: Blue-Green Algae that Transforms Solar Energy into Superfood 188
Sugar: Is it Really "Killing Us Softly With Its Sweet Song?" 190
A Future-Science Strategy for the Production of Superfoods 191
Master Key 20 - Psychic Technology (Psy-Tech) 192
Psi-Microvision: Amazing Psy-Tech Tool for Advancing Conventional Science
  and Technology 193
Remote Viewing as a Key Aspect of Psychic Technology 195
Future-Memory: A Psychic Technology for Predicting the Future 196
Psychic Intuitives and Corporate Mystics in Business 198
Possible Future Applications of Psychic Technology 200

Master Key 21 - Psychic Biology 201
Dark-Field Microscopy and Live Blood Analysis: An Amazing Medical Protocol
  Which Can Interface with Psychic Technology 202
An Amazing Psychic Technology for the Genetic Transformation
  in Plants and Animals 204
Fukushima Disaster Triggers Unprecedented Increase in Airline Pilot
  and Passenger Heart Attacks, Cancers, Radiation Illness Symptoms 206

**Master Key 22 - After-Death Communication: Removing
the "Death Barrier"** 208
Some Early Personal Encounters with Death 208
How the Death of a Loved One Can Become a Golden Opportunity
  for Re-Inventing Ourselves 209
Neurosurgeon Claims to Have Visited the Afterlife: Says, "Heaven is Real" 211

**Master Key 23 - Channeling and Interdimensional Communication
as Key Resources for Consciousness Development** 214
"Conscious Channeling" As a Refined Version of Trance Channeling 215
Channeling as a Valuable Para-Scientific Tool for Accessing Information
  and Guidance from the Higher Realms 215
Personal Experiences with Channeling 216
Why Should Entities in the Higher Dimensions Ever Wish to Communicate
  with Humans? 219

**Master Key 24 - Exopolitics: Will Humans Ever Become Members of the
Extraterrestrial Community?** 220
Exopolitics: A New Protocol for Human and Planetary Evolution 223
Exopolitics and the UFO Phenomenon 224
UFO's: The Overwhelming Evidence 225
UFO Incursions into U.S. and Foreign Nuclear Military Bases 230
Astronaut Edgar Mitchell's Comments on UFOs and ETs 231
Exopolitics: A Wild-Card Solution to Our "Global Dilemma" 233

**Master Key 25 - Quantum Leapfrogging: A Pathway to Transformation
and Enlightened Evolution** 235
Moving into the Future Within the Field of Self-Creation 236
Transformers as "Catalytic Agents" for Raising the Global Social Consciousness 236
Creating Consciousness Fields for Personal, Social and Global Transformation 237
Overcoming Fear and Shifting into a Positive Mindset 239
Consciousness Technologies for Ending War and Achieving World Peace 242
Closing Message from the Author 246
About the Author 247
Bibliography 249

Foreword

I first became aware of the new Master Paradigm *Future-Science Technology* at the Inside Joy Retreat in the fall of 2012 in Sedona, AZ, where I was a guest presenter on the transformation of consciousness through Buddhist prayer and sacred art. It was my honor to be one of the first to learn about Dr. Elliott Maynard's "unique new operational protocol for the human race." Elliott has combined a set of basic principles into a living interactive field which can be applied to all aspects of our lives. In his own words: "The concept is a direct experience of transformation from the past, into present and future-thinking strategies expanded into higher dimensional realms."

During Dr. Maynard's presentation, the beautiful image of a Tibetan prayer wheel spinning in virtual form on the computer screen created a lasting impression on me, as this illustrated how the best of present and future technology can be combined with traditional sacred practices. This brilliant concept blends the best elements of ancient and modern technologies, eastern and western worldviews, and highlights the importance of expanding our traditional ethnic mindsets into new realms of universal cultural understanding.

I am a simple Tibetan monk, teaching Buddhist studies and Tibetan Sand Mandala art. I travel the world to construct Tibetan Sand Mandalas for schools, museums and special events. Prayer wheels and mani-wheels are also part of my sacred Tibetan tradition. Elliott has expanded global access to this one simple, beautiful technique by combining it with other technologies and presenting it in a new way. I think his work in the world of consciousness transformation has just begun, and let us thank him for boldly leading us, with great love and compassion, into the Future-Science World!

– Venerable Tenzin Yignyen

Venerable Tenzin Yignyen was ordained as a monk by His Holiness the Fourteenth Dalai Lama, and entered Namgyal Monastery in Dharmsala, India in 1969. There, he completed studies in monastic rituals and philosophy. In 1985 he received the monastery's highest degree, "Master of Sutra and Tantra" with highest honor, which is equivalent to a Ph.D. degree.

In 1989-1990, Tenzin assisted in the research and translation for the book, *The Wheel of Time Sand Mandala*, in conjunction with the Samaya Foundation in New York City. He has constructed Sand Mandalas in many different venues, including colleges, schools, art museums, Times Square in New York City and the Smithsonian Folk Life Festival in Washington, D.C. He has also created mandalas in Moscow and St. Petersburg, Russia. In 1993, he was invited to the Ganden Thekcheling Monastery in Ulanbaatar, Mongolia where he lectured on the subjects of Kalachakra tantric rituals and mandala construction.

In 1995, Tenzin was selected to teach for three years at Namgyal Monastery's North American Seat in Ithaca, N.Y. (Namgyal monastery is the personal monastery of His Holiness, The Dalai Lama.) Following that assignment, he has remained in the United States, giving instruction in tantric ritual, Buddhist philosophy and Tibetan Buddhist artforms. Teaching has been conducted using both academic environments, and traditional methods which were passed down from the times of the Buddha nearly 2,500 years ago in an unbroken lineage from teacher to student. Tenzin is currently a Visiting Professor at Hobart and William Smith Colleges, and has served as Professor of Tibetan Buddhist studies since 1998. He provides spiritual counseling to groups and individuals, a traditional role of the lama in Buddhist practice.

BRAVE NEW MIND
Living in a Future-Science World

Introduction

Reflections, Confessions, Pratfalls and Youthful Adventures

I remember my grammar school days in the small town of Farmington, Maine. We didn't have computers, cell phones or video games, but outdoor fun abounded. There was a small patch of woods in back of our large house where we could climb young trees, grab the tops and swing back down to the ground. We had a barn to play basketball in; two skating ponds close by and a regular skating rink a short walk away. There was also a nice ski slope with night lighting, so there was no shortage of physical activity. I joined the local youth shooting club, which was located in the basement of the Franklin County Court House. Weekends I would go there with my shooting buddy, Willy McLeay. We would check out our single shot 22 caliber rifles and walk along the railroad tracks, using the occasional discarded cans for target practice.

Willie taught me scale the sides of two-story brick buildings to quietly peek at the rooftop falcon's nests without disturbing them. We also eventually discovered that if you removed the lead bullet from a 22 cartridge, you could insert a short length of dynamite fuse to make a nice little detonator, which worked just fine when pushed through the hole in the top of an aluminum 35mm film can, filled with gunpowder. When detonated at suitable remote locations this created a wonderfully satisfying explosion, complete with a cool mushroom cloud. I remember going into the hardware store and saying, "I would like a pound of galvanized nails…and 20 feet of dynamite fuse." It's something of a miracle we survived our adventures with no one injured, or getting into serious trouble!

My dad was the high school principal, taught English and History and was a consummate outdoorsman. He was captain of the Dartmouth University track team, a national high-jump champion and a dedicated hiker and fisherman. Since both my parents skied, I took lessons from a local Swiss ski instructor and was soon up and running. I got hooked on skiing and entered my first ski races by third grade. I remember my first ski racing experience. I entered the downhill, slalom and ski jumping events. I attacked the course with naïvely reckless abandon, managed to get through the icy gates without falling and somehow beat the reigning fifth-grade champion who burst into tears after his humiliating loss.

Instead of digital books, videogames or computers we had a lovely domed granite library within walking distance from home. During the cold and dark winter nights I would visit this magical sanctuary, check out six or seven books and trudge back home through the snow. I quickly "read out" the kids section of the library, and went on to broaden my literary horizons. My favorite books were pirate stories and any kind of sci-fi, which included the Jules Verne classics as well as the "Tom Swift" boy's series from my Dad's library. I became fast friends with a geek who lived only a couple of houses away. He had a huge stack of comic books, which I proceeded to devour with relish.

Every summer we would drive to the family camp, a turn-of-the-century lodge, built by my grandfather, who was a 32nd degree Mason. The central feature of the expansive living room was a large fireplace set with granite stones, which climbed 20 feet from hearth to roof peak. An antique cast iron wood stove in the kitchen provided a warm retreat for chats with my mother and grandmother on chilly Maine mornings. The camp was set like a jewel into a wild forested property, nestled among tall pines, maples birch and hickory trees, with a large front porch which overlooked the lake. We even had our own beach. Once we arrived at camp I would take my shoes off and go barefoot as much as possible. I became a fishing fanatic at an early age. Each morning I would set out with my fishing rod and walk along the shoreline. Without fail, I would return with a nice string of sunfish, shiners and the occasional catfish for my mother, who would feed them to our two cats. As I grew older I used our canoe or rowboat to fish for perch and bass.

One day one of my dad's fishing buddies came to visit. He brought with him one of the original round rubber diving masks. I was fascinated with this new way to explore the underwater world. I saved up my allowance and soon became the proud owner of my own mask, flippers and snorkel. I was hooked again – this time on exploring the underwater world! By the time I entered high school I had saved up enough money to buy a SCUBA outfit – a single-stage regulator with tank and harness. I was entranced with Jacque Cousteau's book, The Silent World, which I used as a basic diving manual. I remember melting lead ingots on my mother's kitchen stove and pouring the molten lead into a mold to make my own diving weights. My early diving forays into chilly Maine ocean waters were a pioneering effort, as I did not have money for a wet suit, but instead wore a sweatshirt and long johns to ward off the cold 55-60 degrees F water temperatures. I learned how to capture live lobsters on the ocean bottom by removing the snorkel from my belt and distracting a lobster's claws, so I could snatch it up without getting pinched and pop it into my mesh bag.

One fateful summer day I took my friend, Tony, on his first diving expedition. Tony was the president of my high school senior class in Portland. At that time Buoyancy Control Devices (BCD's) had not yet been invented, so we used an inflated inner tube with a mesh bag inside to rest on at the surface and to collect our catch. As we swam out into the sandy cove where lobster boats were anchored, Tony swallowed some water and began thrashing about in panic. In his struggles he ripped a hole in our inner tube float with the lobster gage attached to his wrist. In his desperation he grabbed me and began to pull me under. I was able to calm him down a bit so he released his death-grip hold. Suddenly, he panicked, and then plummeted to the sandy ocean floor about 20 feet below. I followed the trail of bubbles down to the sandy bottom, but was unable to release his tank harness. I released Tony's weight belt then used my diving knife to cut his tank harness and brought him to the surface. I managed to drag him over to a nearby lobster boat, climb aboard and haul him into the stern. I shouted to our friend waiting for us on shore, who called for a doctor. I tried artificial respiration, which we learned in Boy Scouts, but he failed to respond. Shortly thereafter a doctor arrived in a borrowed rowboat. He administered an adrenaline shot directly into Tony's heart, all to no avail. This was my first experience with the death of a close friend.

I remember arriving home and telling my mother, "Tony drowned." She dropped the frying pan she was holding and began to cry. Later, I visited Tony's mother to explain to her what had happened. She just screamed at me: "Why was it you that came back?" I remember attending Tony's funeral and looking at his embalmed body in the casket, thinking that he looked something like a "stuffed fish." At that point I decided that I would remember Tony as he was in life, and vowed to somehow turn this tragic life lesson around. Later Tony's mother gifted me his dictionary, which I used to improve my spelling as I thought my friend Tony, would wish me to do.

In grammar school our teachers asked us to write down what we would most like to do in the world. I remember writing that I would like to leave some kind of mark in the world, and would also like to meet the first aliens to visit Earth. Even then I had a drive to become an ET Ambassador, though I also remember how the thought of such an encounter was exciting and terrifying at the same time.

At that time I never imagined how bizarre and convoluted my life-journey would turn out to be. As a college sophomore this colorful life odyssey included a special training for the 1960 Olympics at Yale University along with a select cadre of top-flight swimmers from Harvard, Princeton, Yale, the U.S. Navy, plus national champions from the U.S., Israel and Cuba. I arrived at the Yale University training camp a virtual unknown (the pool at my alma mater, Washington and Lee University was a half-yard short of regulation length). We slept on cots in the Yale rowing tank rooms. Each morning we were awakened by legendary Olympic coach, Bob Kipputh, who took fiendish delight in firing his starter's pistol in the cavernous rowing tanks to roust us out of bed at 7am sharp. This formidable elfin character would appear, dressed in leotard pants and sweat shirt top, small pot belly and a huge wooden staff in hand, which he used to coordinate our morning exercises like some Gandolphian boot camp drill instructor.

I trained diligently, and subsequently managed to break and hold an American record for about 20 minutes during the 200m breaststroke trials for the National AAU (Amateur Athletic Union) Championships in Cleveland, Ohio. We then went on to the Olympic Trials in Detroit, where I placed eighth in the finals of the 100m breaststroke, and then ninth in the 200m breaststroke event (Only the first three winners get on the U. S. Olympic Team). Realizing I had failed to make the finals in the 200m race, I went back to the locker room, woofed down a bag of potato chips and washed them down with a soda. Then…I heard my name called on the PA system… Arrrgh! I had tied for eighth place in the finals and a tiebreaker swim-off would be held shortly. I guess my angel du jour was looking out for me, since I won the swim-off, with my training partners cheering me on. Thus ended my erstwhile Olympic career.

Looking back on my life from the point of this writing I realize that, even in my wildest imagination, I could never have dreamed I would go on to realize so many of my childhood visions and dreams. Despite some deeply painful episodes and major disappointments in my life, I never gave up. I always tried to make the best out of a bad situation and somehow find a way to make lemonade out of any lemons I was dealt. As a college freshman I remember one of my biology teachers telling us, "The amount of satisfaction gained by any organism is directly proportional to the amount of struggling it has to do to get there." Tragically sad, but oh so true!!

One Tibetan sage is purported to have called the physical realm we live in "the trashcan universe." He also said something to the effect that, "Humans are not naturally stupid; they go to school to get that way." After twelve years in higher educational institutions and gaining degrees and graduate experience in Zoology, Marine Sciences, Coral Reef Ecology, Phytoplankton Ecology and Rainforest Ecology, I suppose I should have ended up terminally stupid. Thanks to the school of hard knocks (the ultimate reality show) I learned to work the system and creatively apply these life lessons in my own way. There is indeed a balance for all things!

I included this personal introduction to provide you readers with some of my background information. Where am I going with all this? During the past few years I can say I have finally been able to gain enough wisdom, courage and enlightenment that I have experienced periods of happiness and well-being that I never before dreamed possible. My personal mission in life is to share some of the amazing things I have learned over my lifetime, and provide a potpourri of cutting edge information so you can research items of interest and judge for yourselves. I also fully understand that most of the gifts I have received are "offers we can't refuse," since it is necessary to share these gifts received with others, so we can keep consciously evolving. I have attempted to shed new light on old subjects, bring into the forefront orphaned and cutting edge technologies and to create positive new ways to make this world a better place. This way, we can work together to support the survival and success of the human race, as well as our beautiful home world, Earth. So…Welcome to my Quantum Theater of Dreams. The footlights are dimming. The audience waits in hushed expectation. The curtain is beginning to rise. Are you ready to expand your own brave new mind? If so…prepare for liftoff. Welcome to the Future. Your adventure in quantum-field thinking is about to begin!

What is Future-Science Technology?

Future-Science Technology is the quantum-field science for the future – a Master Plan for Humans and Planet Earth. The twenty-five keys make up the basic fabric of Future-Science Technology, with each key representing a different approach for creating connections between left-brained logical thinking and right-brained intuitive thinking (The Quantum Field). It is this logical left-brained perspective that has dominated western thinking since the invention of the Gutenberg Press. This invention ushered in the age of mass media.

Mainstream scientific paradigms are based on hard facts, generated from a logical thinking perspective. The basis for "hard science" dictates we believe nothing that we cannot prove to be true through measurements, replication and mathematical validation. The overemphasis on facts, formulations and replications has led establishment scientists and world leaders to support institutions, which focus on compartmentalization, overspecialization and compilation of vast amounts of information. This linear approach dominates most of our existing scientific, governmental and military organizations, since the bulk of research funding is focused on supporting this kind of approach.

Historically speaking, what we think of as "modern technology," has only existed for merely an eye-blink on the greater historical timescale – since the beginning of the industrial revolution in the late 18th Century. In contrast, right-brained approaches to science, medicine and the arts have been essential elements of Native American and Eastern philosophies for thousands of years. The great scientists, artists, musicians, mystics and visionaries of the European Renaissance were only able to break through the restrictive logic – based paradigms of their times because their creative achievements in science, art and music became so obviously brilliant that it became "fashionable" for the powers that be to support them and thus gain an extra bit of fame for themselves. What was created was essentially "an offer they could not refuse." This same creative persistence for working outside hard-science protocols is alive and well today. Otherwise, we would not have Microsoft, Apple, Yahoo, Facebook, Tesla Motors and countless other paradigm breakers. Remember that most of the key innovators of these new paradigm companies worked with little support or encouragement, often from garages and against daunting resistance from established institutions. What did they all have in common? They believed in themselves and in their innovative creations. They pushed on against all odds. They defied establishment logic to "search" in deference to logical "research" and endless data-churning. Their success became so glaringly apparent it became uncomfortable for establishment scientists to continue ignoring their work. Success speaks for itself. Genius speaks for itself. This forces naysayers to shut up and "get with the program," because their positions of resistance no longer benefit them. These innovative concepts were only successful because modern geniuses transformed their visions into practical working models such as the basic hardware and software innovations of Microsoft and Apple Computers. These systems are now so deeply imbedded in our lives that we simply accept them without thinking.

Neo-Renaissance thinking embodies the concept of learning to use both hemispheres of the brain with regard to man's relationship to his environment and to human society in general. What this accomplishes is to go back on the evolutionary time-scale to the point where the bi-hemispherical structure of the human brain was created. At that time the human brain was shaped by the major drives for survival and success within the global environment.

When humans made the shift from hunter-gatherers into agriculture, a new mindset emerged. This new perspective put man more in touch with his external environment, as he abandoned his former nomadic ways and became engaged in the responsibilities of planting, harvesting and storing crops for the winter. This new routine encouraged the formation of permanent villages and also allowed more time for the sharing of wisdom, development of new trades and for the expansion of personal consciousness.

In contrast to this gradual shift experienced by our ancient relatives, our lives in the present have been inundated by lightning-paced developments in computer technologies, communications and the global internet. These recent technology shifts have forced our personal perspective to expand exponentially. Thus, the worldview of our parents' and grandparents' generations has grown from a few square miles to encompass the entire planet and extend out to embrace the far galactic reaches. Recent developments in space sciences and related digital technologies, have thus served to extend the scope of human consciousness to reach out across vast interstellar distances to even the most distant galaxies. This quantum expansion and evolution of the human external perspective has been reflected back in corresponding shifts, which have profoundly evolved the fundamental nature of human consciousness itself.

With the advent of Earth-orbital satellite technology, lunar, planetary and interstellar probes, and space observatories like the International Space Station (ISS), the Russian Space Station, Mir, and the Hubble Telescope, the Planetary Consciousness Field (global brain) has undergone a corresponding expansion in its self-awareness, which now embraces the vast reaches of interstellar space.

Future-Science Technology represents a new operational protocol for the human race. It implies a major shift from a past "rear-view mirror" perspective into a present and future-thinking perspective. This master paradigm embodies the key components of both evolution and transformation as "catalytic change-agents" and implies a shift from local and national perspectives into an entirely new spherical consciousness framework. This new consciousness framework is thus truly global, as it highlights the future of humans and their interrelationships with Planet Earth. It also acknowledges the greater extraterrestrial community, and embraces the hope that the human race will soon become enlightened enough to be invited to participate with extraterrestrials in meaningful exchanges of wisdom, trade, advanced technology and cultural exchange.

The time has arrived in Earth's evolutionary development for the new thinkers of all nations to make the quantum shift from a regional/national perspective, to a fresh new planetary and intergalactic consciousness. Our technology has reached the point where we can either choose to enhance – or continue to systematically destroy – the delicate balance of the natural forces, which sustain our global biosphere. If mankind can fully understand the critical priorities of ecological preservation, restoration and enhancement as prerequisites for the survival and success of our now and future generations, humanity will have achieved a profoundly important transition into an entirely new and unprecedented phase of its evolutionary journey. This is a journey of quantum transformation into a brave new mindset. This is the wondrous journey of Future Man – Homo novus. Your adventure has just begun!

Part I
Working with the Quantum Field

> "Tuning into a fully developed Master Field,
> where all of the ordinary evolutionary processes have already taken place,
> permits those changes to be magnified and quickened
> or, in effect, lived into that system without its having to pass through
> certain elements of the processes associated with the individual to evolve.
> Therefore, as I have indicated, the spiritual adept is a unique mechanism in Nature
> provided for the sake of the spiritual and altogether human evolution of human beings,
> as well as the transformation and evolution of all beings,
> and all processes that exist in the cosmos."
>
> Da Free John, 1978 – *The Enlightenment of the Whole Body*

Linear vs Quantum Thinking from a Future-Science Technology Perspective

Future-Science Technology involves the expansion of our traditional linear social consciousness into an advanced modality of non-linear thinking, so we can consciously function within the quantum realm. Considering the recent acceleration of human consciousness, combined with Planet Earth's ongoing evolutionary shifts, it is no longer necessary to sit in a cave and meditate for 30 years to become "enlightened." We simply need to open our consciousness to the possibilities of becoming interdimensional, and to accessing the quantum field. We have the ability to shift into this advanced modality at will. This is accomplished by simply becoming aware of the *shift* from linear mode to a non-linear (quantum) thinking modality. We can then continue to reinforce this concept by focusing and "living into" this new awareness. This linear to quantum shift might best be compared to working with audio or video tapes as opposed to CD's and DVD's. For example, with tapes it is necessary to rewind or advance the tape to access a certain data point on the track. This can be compared with linear thinking. With the CD or DVD, you simply go directly to that point on the disk where the desired information is located. This is analogous to working in the quantum field. This is a simple analogy for accessing the quantum information field, where data can be acquired instantly and non-locally. Physicists have sometimes referred to this quantum information field as "the whispering pond," as it contains an infinite sea of data. Since this quantum field functions holographically, with data existing outside the linear concepts of time and space, information can be accessed directly and instantly and is perfectly relevant to any specific situation or circumstance. By expanding this metaphor to include the idea of interdimensionality, it is easier to grasp the complete picture of this concept.

As children, we learn by continually adding new words to our vocabulary, and adding new facts and experiences to our wisdom banks. In keeping with standard linear thinking modalities via books, digital media and traditional learning institutions, we are taught to file these facts in chronological order, and then regurgitate them on demand in tests and essays. As we grow older, we find ourselves overburdened with the sheer numbers of experiences and facts, so we learn to compartmentalize this data – much like we create files in a file cabinet, arrange books on shelves, or set up files and folders on a computer.

By contrast, whenever we engage in physical or creative activities, we rely mainly on our non-linear thinking modality. This is true when we ride a bicycle, drive a car or engage in art, music, martial arts or dance. When we are engaged in these activities our non-linear intuitive faculties guide us. This is because it is impossible to respond fast enough in the linear modality to make the precise corrections needed by a skier or surfer who makes countless physical and mental adjustments every second.

Since non-linear thinking is perfectly natural in our more primitive or childlike mindsets, learning to operate in the quantum field thinking modality starts when we "switch on" the awareness that this state exists, and then cultivate an awareness of *when* we are operating non-linearly. Once we start to recognize this new sense of awareness, we can teach ourselves how to operate more efficiently and more often within the quantum field modality.

Collective Consciousness:
A Powerful Resource for Igniting Positive Change

The power of collective consciousness has tremendous potential for effecting meaningful and lasting change in the social consciousness, as well as in the global biosphere. In a study by Dr. Mitchell Krucoff, head of Duke University's cardiovascular program, a group of patients complaining of chest pain was divided into two groups of equal size. One group was "prayed for" and the other group was "not prayed for." The request for prayer was carried out by a global prayer group, which had already been established on the internet. This interdenominational group included Buddhists from Nepal and Tibet, Hindus in India, Carmelite Nuns from Baltimore, a prayer group from Jerusalem, Unity Church members and a group of fundamentalist Protestants from North Carolina.

Results from the initial study were surprising! Patients in the designated "prayed for" group were observed to have *from 50-100 percent fewer side effects* than the control group that was not prayed for. The initial study was expanded to include twelve other major U.S. hospitals. An interesting sidelight of this study was that these positive research results were supported by dozens of other studies, which used non-human subjects [Using non-human subjects quashes a major argument by traditional scientists, who often cite the possibility of "hidden biases" which might surface in research results, due to conscious interference by human patients]. Using animals, plants, bacteria and germinating seeds eliminate this type of "human bias factor." Studies with non-human subjects included healing rates of surgical wounds in animals, bacterial replication rates, growth of fungus colonies in Petri dishes, growth rates of seedlings under controlled laboratory conditions and the rates of certain biochemical reactions in test tubes. Since experiments on non-human subjects can be conducted under precise laboratory conditions, this effectively eliminates the major objections cited by skeptics of human studies in areas of consciousness research.

One fascinating aspect of this research is that the non-human studies highlight the "nonlocal consciousness effects" which can operate across a broad spectrum of natural phenomena to yield measurable effects which range from the sub-microscopic realm (atomic and molecular scales) to the levels of bacteria and larger single-celled organism such as protozoa. Similar effects have been observed with multicellular organisms such as invertebrates and vertebrates. In the words of Larry Dossey, M.D., "…this lineage, this so-called concatenation or coming-together of effects unifying these vastly different domains of nature, is one of the most compelling aspects of the field." The significance of this unifying concatenation is that it represents a highly valued aspect of validating scientific theory and suggests we are dealing with a unifying principle that is embedded everywhere throughout the patterns of nature.

The foregoing studies focused on two main biological processes: healing and fertility. Healing research involved wound repair and the healing of specific conditions like AIDS and coronary artery disease. Fertility research involved humans, non-humans, growth rates of plants and germination rates of seeds. According to Dr. Dossey, "The secular scientific and religious sides of life cannot be kept in separate boxes." Notable scholars such as Alfred North Whitehead have expressed similar sentiments regarding the inseparability of science and religion: "It is no exaggeration to say that the future course of history depends on the decisions made by this generation as to the relationship between science and religion." Ralph Waldo Emerson viewed this concept from a different viewpoint: "The religion that is afraid of science dishonors God, and commits suicide." This sentiment was also echoed by Albert Einstein, who stated: "Science without religion is lame. Religion without science is blind."

This research highlights the importance of religious tolerance and interfaith cooperation, since, regardless of one's faith, the end results appear to be universal. Studies like this seem to provide a measure of hope for our presently troubled society and our environmentally stressed biosphere. They bring into the forefront that the concepts of love, empathy and compassion have the power to change the state of living organisms and the natural world (Dossey, 2002).

Internet-Based Torsionic Meditation: A Mechanism for Shifting the Global Consciousness

Torsion field generators transmute spinning fields of negative chi energy into positive chi energy. This serves to release stress so our body can begin to restore itself. By lowering stress our immune system is enhanced and our pineal gland can open up to activate our 3 to 12 strand DNA. Since DNA *is* essentially structured water it can be influenced by our thoughts and feelings. According to quantum field entanglement theory, all parts of the whole are in all places at all times. Thus, if we tune into these same frequencies, which are based on sacred geometry and solfeggio frequencies, we can tap into a version of ourselves that was healed in a different place or time.

Since vibrational frequency is the very essence of consciousness technology, Yogis and ascended masters are naturally in tune with this consciousness energy field. Traditionally, it took 30 years of daily training to achieve this ability. By contrast, torsion field energy can accelerate this process from *years* into *months*. Since this technology is based on pure consciousness it is very powerful. When we tune into the proper vibrational frequency we can tap into the universal collective consciousness. Prayer, meditation and relaxation are thus important in our lives, since we can only tune into these higher dimensions when our body is in the right state of mind. Since our bodies are essentially *scalar wave energy receivers* we have the ability to tap into the quantum energy field. This allows us to manifest our desires.

According to Russian researchers spiritual teachers, Nicholas and Helen Roerich, laid the basic groundwork for torsionic meditation.

From a Future-Science perspective, Buddhist prayer wheels and computer hard drives can both be regarded as "generators" since they have the capability to propagate faster-than-light torsionic emissions, which are said to be capable of carrying messages over interstellar distances.

In the 1970's a U.S. Air Force installation in Thailand donated some obsolete computer equipment to a local community of Buddhist monks, who began experimenting with computer hard drives, attempting to use them as modern technical analogs to Buddhist prayer wheels. Since the drives are mass-produced, they offer the advantages of being relatively inexpensive, yet are created within precise manufacturing specifications. Computer hard drives are essentially finely-tuned spinning fractal multi-magnets. With millions of computers being connected to the global internet, the potential exists for a planetary-scale phased torsionic antenna system that could theoretically be effective for purposes of mass-meditations and global paradigm shifting.

∞

A New Format for Future-Science Paradigms

> Traditional "futurists" project past linear trends into the present and future. However, a new and unique breed of futurist is emerging in human society. These pioneering researchers work within the context of quantum consciousness, with the goal of accelerating positive global transformation. These change-masters have the ability to select and empower the most practical and positive future probabilities. They are actually creating new evolutionary futures for the human race."
>
> *Elliott Maynard 2009*

Historically, mainstream science has devoted a major part of its resources toward protectionism and self-perpetuation. This has resulted in an overriding tendency to emphasize the importance of "re-search" instead of "search." Periodically, these institutions have been subjected to such embarrassing glimpses of the obvious, that they have been forced to modify their existing paradigm. They needed to make the appropriate shifts to accommodate new scientific systems and devices that existed outside their restrictive guidelines. Classical examples of institutional resistance were encountered by pioneering scientists who dared to challenge the antiquated religious and scientific institutions of the times, because their heretical ideas conflicted with established thinking. Modern examples of hauntingly similar disputes can be found in the irrational reactions generated by government and university scientists against pioneering researchers in the field of cold fusion, or the various excursions into areas of zero-point energy and quantum physics, where established physical laws break down, and the distinction between science and consciousness is blurred.

Old Paradigms: Examples of the Industrial-Age Mindset

Prior to the year 2000, paradigms typically took the form of operational guidelines for thought and action. Leading-edge 21st Century thinkers tend to regard these old paradigms as restrictive thought patterns and societal hangovers from the Industrial Age. Scientists and leading-edge thinkers of the past, presenting ideas that were outside these restrictive thought patterns often suffered public ridicule, severe punishment or even death.

Although the following two quotations were made by respected scientific and military authorities of their times, a generation or two later, these "truths" now appear as "caricatures" when viewed from our present perspectives. From our present-day perspective, they now appear stranger than fiction!

> "X-rays are a hoax.
> Aircraft flight is impossible.
> Radio has no future."
> *Lord Kelvin – Physicist and Mathematician*
>
> • • •
>
> "The Bomb will never go off,
> and I speak as an expert on explosives."
> *Admiral William. Leahy – U.S. Atomic Energy Project*
> *Physicist and Mathematician (1804-1907)*

New Future-Science Paradigms as Triggers for Consciousness Transformation and Human Re-Evolution

The transformative paradigms of Future-Science Technology offer fresh pathways for evolution and social transformation, when compared with the restrictive paradigms of the past. Through creative new applications of computer technology advanced paradigm models can be designed which can function independently as "self-aware concepts," since they can be created with a built-in capability for self-organization and evolution. They can also be mentally programmed with the abilities to grow, adapt, and evolve, so they provide the most appropriate support for the social and ecological conditions of present-day life.

Unlike paradigms of the past, which were severely rigid and constrictive, Future-Science paradigms are designed with guidelines, which are both flexible and evolvable. These guidelines form a matrix for creative energies to enrich mental and spiritual growth. They also encourage intellectual and cultural diversity, and encourage us to develop innovative solutions for the social, economic and environmental problems we face as present and future citizens of Planet Earth.

Embedded within the conceptual matrix of the Future-Science paradigm model is a wealth of new solutions, technologies, concepts and programs. Each proposed decision and action is thus evaluated in terms of its potential impact on present and future generations. Examples of this include simplistic concepts such as "creating win-win solutions," "giving back more than you take" and "leaving a place a little nicer than you found it." The creation and focus of intent on such positive thought matrices, when multiplied exponentially by millions of brave new human minds on Earth, can serve to create a bonanza of bright future scenarios for the global biosphere and for generations yet to come. Here are two examples of statements, which reflect the essence of new co-evolutionary paradigms:

> "If you wish to make apple pie, you must first invent the Universe."
> *Carl Sagan – 1934-1996*

> "From the perspective of Future-Science Technology,
> a website can be described as
> a self-aware communications interface
> between humans and the planetary consciousness field."
> *Elliott Maynard, 2003 – Transforming the Global Biosphere:
> Twelve Futuristic Strategies*

The Breakdown of Conventional Scientific Thinking: Re-Inventing Science to Fit the Times

In many areas of global ecology, mainstream science has failed to provide workable solutions for the problems caused by human activities and the forces of Nature on the global biosphere. Our technology appears to have overrun our ability to use it intelligently, and science has essentially "hit the wall." Conventional twentieth-century thinking simply falls short when it comes to generating innovative and practical solutions for dealing with the major problems facing the future of Earth and the human race.

In response to this dilemma of unprecedented proportions, the author proposes an entirely new method for designing scientific paradigms, which integrates conventional science with alternative science, combines ancient and modern technologies and incorporates ecologically appropriate technologies which lie outside the present boundaries of conventional science. The methodology for this new-paradigm thinking has existed for several decades, but has been intentionally pushed aside (or simply ignored) by establishment scientists and corporate entities, whose main interest has been to maintain the bottom line at all costs, rather than to accept the challenges and responsibilities of actively moving into the Future.

"Brave New Mind"
A New Operating System for the Future

The key to creating new evolutionary pathways into the third millennium requires a concerted effort by world leaders to encourage the development of new innovative scientific systems, and to support their development and integration into practical programs. This renewed spirit of support from all sectors has become an "evolutionary imperative" if we are to survive and thrive during the coming social and environmental shifts. An inspiring new perspective is a prerequisite for guiding us all to produce environmentally cost-effective solutions, which can be practically integrated into the fabric of society. If this Future-Science perspective can be actively embraced and integrated into mainstream consciousness, resistance from formerly established forces will simply melt away. This way, everyone benefits – personally, institutionally and economically with a new sense of enthusiasm and enlightened purpose. For such a major shift to become reality, this shift must begin in the hearts, minds and spirits of individuals, and expand exponentially to become the common social imperative for everyone on Earth.

Global society must make a profound shift from its typical dysfunctional fear/ego-based mindsets (*Kill or be killed*, *me-first*, *environmental plundering* and *over-consumption*) into a new mindset which incorporates *self-respect, self-responsibility, self-motivation* and *concern for the welfare of others*). This also includes a heightened sense of *awareness, respect* and *nurturing of our planetary resources*. Within a new framework of synergistic support and international cooperation, egocentric barriers will more easily dissolve, and the defensive attitudes of fear and insecurity, which have dominated human society over the past millennia, can be left behind. For this transformation to occur both conventional and alternative scientists must put aside their egotistical, territorial and defensive attitudes. By forming a united front based on enthusiasm and cooperation, these two groups combined, represent a powerful intellectual resource for the continued survival, success and prosperity of our scientific and industrial institutions. The focus of this new thinking should be to demonstrate that it *is* possible for alternative scientists to establish a synergistic relationship with traditional scientific establishments, and to work to generate environmentally appropriate new technologies, which combine the best of all worlds.

New Pathways for Human Enlightenment and Biospherical Restoration

New of win-win transition initiatives need to be created so existing establishment interests can discover profitable ways to save face and prosper even more efficiently, by empowering their intellectual resources and increasing their cash flow and profit margins. Alternative scientific concepts and applications, on the other hand, need to be grounded into working models and practical applications, which demonstrate indisputable results. This approach allows for sustained forward progress and productivity, with theoretical principles and detailed scientific validation left to be defined by scholars at a later date. Economics is a basic driving force for scientific and technology development, as it can facilitate the integration of advanced scientific principles into mainstream social functionality. Cooperative investing by concerned individuals and corporate angels would further accelerate the growth and integration of advanced technology, as this would serve to stimulate the viral blossoming of profitable and ecologically appropriate new technologies for the 21st Century and beyond.

Within the context of Future-Science Technology, the posturing, territoriality and survival-of-the-fittest mindsets of corporate moguls and ivory-tower scientists would ultimately yield to the birth of practical applications, directed toward improving the quality of human life, within the framework of ecological sustainability. *Search*, in the form of disciplined discovery would replace the traditional academic *re-search* treadmills and a new and enlightened sense of synergistic creativity would prevail. The massive hemorrhaging of public tax dollars, which continue to fund destruction-focused global military activities, could be re-directed into the formation of a new global environmental task force. This organization would focus on conflict prevention and resolution, as well as the protection of our global commons. This would also include disaster relief, implementing environmental clean-up and restoration programs, shepherding forest and ocean fishery resources and providing technology and manpower to manage intensive food production systems. Such a new eco-military program would require an elite cadre of leaders, who would participate in meaningful programs for planetary protection. During their terms of service, enlistees would learn valuable career skills, which would continue to be relevant in civilian workplaces after their eco-military service.

By applying these enlightened applications of third-millennium thought and action, global wilderness preserves and city forest parks can be developed, using genetically enhanced trees, selected for their rapid growth, air-cleaning capabilities and natural beauty. Hatcheries and nurseries can be established to re-seed the world's oceans with fast-growing, disease-resistant fish and shellfish, and lost or damaged corals can be farmed and re-planted on the major coral reef ecosystems of the world. Treated human and animal waste can be utilized to increase the fertility of crops worldwide and to fertilize blooms of beneficial plankton in marine environments everywhere. This would serve to enrich the ocean pastures and capture carbon dioxide from the atmosphere while reducing pollution in the world's most densely populated coastal regions.

National programs can be set up to rejuvenate, monitor, and manage ocean fisheries, using satellite imaging technology and remote transmitting ocean buoys. These programs would serve to pinpoint environmental problems and essentially monitor the "health" of the global biosphere on a real-time basis. Cooperative programs, established and funded by government, corporate and private sectors, could assist in managing wild game preserves, encourage eco-tourism and protect endangered species, with the ultimate objective of re-establishing a healthy balance between humans and the planetary biosphere.

PART II
THE CORE PARADIGMS OF FUTURE-SCIENCE TECHNOLOGY

MASTER KEY 1
Supersensonics: A Tool for Evolving Human Potential

> "Your higher self is a collective body of energy.
> It is vast and ageless; it is everything that you have ever been,
> stretching back to the edge of infinity.
> Within it is all the knowledge that you will ever need and through it
> you can experience a limitless understanding of yourself and the physical plane,
> as well as of the unseen dimensions that lie close at hand.
> Your higher self is sustained by an intrinsic energy that is even bigger than itself.
> This sustaining energy we will call The Force, or The Living Spirit."
> *Stuart Wilde, 1984 – The Force*

Supersensonics is the science of the development, enhancement and practical applications of our extra-sensory capabilities through simple meditation, yoga and biofeedback programs. This concept was developed by Dr. Christopher Hills, founder and president of the University of the Trees in Boulder Creek, California in the 1970's (Hills, 1979). From a biological perspective, Supersensonics is no more remarkable than the more obvious senses of hearing, vision, taste, touch and smell. These supersonic senses function so predominantly in non-human species that they are fundamental to their basic survival and daily lives. Thus, the wondrous migrations of fishes, whales, sea turtles and birds across thousands of miles of trackless ocean; the incredible homing instincts of pigeons, dogs and cats; and the uncanny accuracy of insects in locating their mates all demonstrate abilities which far surpass our physical senses.

Prior to "modern civilization" primitive man's supersensonic abilities functioned as an extension of his physical senses and were thus key factors for his survival and evolution. Now, through specialized training, the natural human supersensonic faculties can be awakened and developed to levels of sensitivity that exceed the range of state-of-the-art electronic instrumentation. Since Supersensonics operates in a non-linear (quantum) modality that exists outside the realms of conventional Newtonian physics, it provides a simple means to break through the restrictions imposed by our conventional scientific, educational and technological institutions. Super- sensonics might also be considered to be the ultimate approach to biofeedback, since it offers exciting possibilities humans to function as "biological computers" – supersensitive instruments, capable of operating in a quantum format far more complex than our most sophisticated computers. Once the human consciousness is expanded to encompass the quantum field of possibilities, our brave new mind becomes activated.

Nuclear Evolution and the Antakarana System

Nuclear Evolution encompasses the development, realization and expression of an individual's maximum potential as encoded in the genetic constitution of every cell in the physical body. Expressions of this potential are not limited to the physical body, but also include extra-physical consciousness centers such as the chakras (Hills, 1975).

A second important aspect of Nuclear Evolution involves the expansion and refinement of the nuclear center (core self). This process also implies that once we reach a certain level of proficiency in a given area, we can experience quantum leaps in consciousness and move into increasingly higher vibrational states, where different sets of rules and higher requirements for the use of wisdom, power and integrity prevail.

Nuclear Evolution is an advanced sociological concept when compared with mainstream social consciousness. It embodies a communications structure (the nodal network), which operates on the principles of *vibration*, *harmonics* and *resonance*. This nodal network utilizes the natural antenna system (*Antakarana system*) that exists in all life forms. Through specialized training it is possible for an individual to learn to use and expand this system, both in terms of *structure* and *function*.

A geographical analog of the Antakarana system exists on a planetary scale as the Earth Energy Grid. This grid can be physically activated at specific geographical locations by applying universal architectural principles, designed to be in conjunction with the Earth Energy Lines (Ley Lines). The Golden Section of the Parthenon and the precise mathematical alignments of the Giza and Mayan pyramids provide notable examples. By expanding the basic idea of a global antenna system, enlightened governmental and spiritual organizations throughout history have used these key power centers to expand, sustain and protect their spheres of power and influence. The great civilizations of the past also knew how to apply these positive energies to quicken their cultural development and enlightenment (Hills, op cit.).

The Art and Science of Dowsing:
A Key Aspect of Supersensonics

From a historical perspective, dowsing with forked tree branches, rods and pendulums was practiced in ancient Rome, Egypt and China. The practice was also well-known throughout Europe where it was used to locate mineral deposits as early as the 1500's. Dowsing eventually attracted the attention of the Church to the point where it was officially denounced by Martin Luther as "the work of the devil." Despite concerted opposition from religious leaders, this occult science continued to attract widespread attention, especially in France during the 17th century.

Prior to the 1890's dowsing was mainly used to locate water or metal ores, but in the 1930's "adverse energies" became a major focus of the dowsing community. This topic expanded dowsing technology to include finding missing persons and also established the science of map dowsing as a legitimate practice. Later, it became commonplace for "water diviners" to dowse maps to locate prospective drilling sites, and then travel physically to these sites to mark the most favorable locations. Since the 1950's dowsing has been used extensively in the fields of geology, mining, engineering, law, medicine and various other disciplines

Although mankind has been dowsing in one form or another for thousands of years, no one really understands how the phenomenon works. Essentially, it is thought that the dowsing tools amplify and connect the operator's consciousness field with the quantum information field, so veins of water or mineral deposits can be detected, and even quantified with regard to amount and depth of the deposit in question. Dowsing techniques are presently used to locate missing persons, find buried water and electrical lines, and to locate buried treasure, oil and mineral deposits, archaeological ruins and historical artifacts. An excellent overview of dowsing, its applications and extensive historical documentation can be found in Christopher Bird's book, *The Divining Hand: The 500-Year-Old Mystery of Dowsing* (Bird, 1993).

My own introduction to dowsing took place as a teen-ager growing up in Central Maine, when I discovered my dad and uncle attempting to locate an underground water pipe at the family camp, using a forked branch cut from an apple tree. I was immediately fascinated with this process and enthusiastically joined in this new adventure. I still remember the moment. I walked over the buried pipe, holding the forked stick in my hands. The forked stick turned down so abruptly toward the buried pipe, that it nearly twisted the bark off the ends of the branch I was holding. During the years that followed my fascination with dowsing continued. I became associated with the United States Psychotronics Association and later served on their Advisory Board. This was during the time when leading-edge scientists such as Dr. Andria Puharich, Dr. Andrew Michrowski, Chris Bird, Dr. Tom Bearden, Dr. Tom Valone, Bob Beck, Marcel Vogel and other pioneers in the field of radionics, ELF electronics, alternative energy and healing accomplished some of their most significant work. Within this group of maverick scientists I soon discovered that nearly everyone was either an accomplished dowser or used a dowsing-related technique most often called kinesiology (i.e. muscle testing) in their research and daily lives. In fact, dowsing had become so much an integral part of our daily lives that the members of this group did not consider it anything out of the ordinary.

Dowsing: An Ancient Science for Amplifying the Intuitive Senses and Interfacing with the Quantum Information Field

Pendulums, L-Rods, and Muscle Testing: Throughout history many specialized dowsing tools have been developed. The most commonly used devices are pendulums and L-rods. Pendulums consist of a short jewelry chain with a weight attached to the end – usually a metal, crystal or mineral bob. The operator holds one end of the chain and allows the weight at the other end to swing freely. The operator then clears their mind of mental static and quietly asks a question. The pendulum will begin to rotate as if it has a life of its own – in either a clockwise or counterclockwise direction to give a corresponding *yes* or *no* answer (the direction for *yes* or *no* having been previously set by the *intent* of the operator). Since pendulums are compact, they can be carried in your pocket or purse and used without attracting unwanted attention. They are routinely employed by healers for balancing chakras, diagnosing disease conditions and determining nutritional imbalances or the suitability of specific nutritional supplements. Some Feng Shui practitioners also use pendulums to clear negative energies from homes, offices and living spaces.

Many naturopathic physicians and doctors use a different method of dowsing in their diagnosis, which they call muscle testing, or *kinesiology*. With this procedure, the doctor has the patient extend their hand at arm's length and asks specific questions or holds certain nutrient samples, while attempting to push the patient's arm down to determine the amount of muscle resistance. Depending on the degree of resistance, the practitioner can check various areas of the body and determine which nutrients are the most suitable for this particular patient. In a variation of this procedure, the doctor simply uses one hand to bend his or her own thumb inward to determine the degree of resistance. Although the technique may differ with each practitioner, the process is essentially an extension of basic dowsing technology, and has been quietly accepted as an established protocol by many alternative physicians throughout the world.

Dowsing with L-Rods: Although I often carry a pendulum with me when I travel, I have found that L-Rods offer many advantages as a useful tool for accessing information from the higher realms. Accordingly, I always carry a set in my computer case and in the glove box of my car.

L-rods are simple and inexpensive to make. They are made by taking a bronze welding rod (available from most hardware stores) and cutting it in half to produce two equal lengths about 16" long. Each rod is then bent to form the letter "L" in such a way that the longer section is about 12" long, and the leg of the "L" is approximately 4" long. [These dimensions are simply guidelines for convenience]. If welding rods are not available, stiff wire, or a metal coat hanger wire works equally well.

The rods are held loosely by the short leg of the L, with elbows bent at 90 degrees so the longer sections of the rods point forward. The long section of both rods should be facing directly in front of the body so they are parallel, and spaced about two inches apart. Prior to dowsing, it is important that the operator fix firmly in mind the intent that a *yes* or positive answer be indicated by the rods swinging open in an arc to a position of 180 degrees so they point directly to the left and right of the body midline. A *no* answer is indicated by the rods swinging back so they swing together from the out front and parallel position to cross each other. Once the basic intention is set in the mind of the operator, he or she enters into a contemplative state and asks specific questions, either out loud or silently. It has been shown that nearly anyone can become an expert dowser. With a little practice the rods will respond quickly and decisively. If the operator is open to allowing the rods to swing freely he or she can also program them to indicate the *degree* of yes or no. [Occasionally, the rods will swing back and forth from open to crossed position, indicating that the question posed could be answered either way]. This simple answering function of the basic dowsing technique can provide simple answers to basic questions such as, "Should I attend a certain event or meeting?" or "Would this book be better than that one?" Being able to determine degrees of *yes* or *no* is useful, since it can save time and money and increase our efficiency and productivity. Dowsing is also a handy tool for setting priorities in our personal and professional lives.

L-rods are most often used to determine the direction and location of a desired target. Professional dowsers most often use this method when they begin working a major site. In this case the operator simply holds the L-rods parallel and facing forward and asks, for example, "Which way is the most promising area for a water well? The operator then slowly turns in a 360-degree circle and the rods will indicate the best direction to go. Experienced dowsers normally use this method from several different points on a map or in the field in order to triangulate the specific location of a well site or other target.

L-rods can also function like a GPS (Global Positioning System) or compass, and are thus useful as a back-up survival tool in the wilderness. In this case the operator simply holds the rods out in front of him and asks, "Which way is North?" or "Which way is base camp?" while slowly turning in a complete circle.

In modern times dowsing is often used to locate buried underground pipes and electrical or telephone conduits. Although it is not common knowledge, many electrical and telephone utility workers carry a set of dowsing rods in their vehicles and use them frequently to supplement their electronic metal detectors.

In addition to the more conventional applications of dowsing, many other useful applications have been well documented. Examples of such non-conventional applications include: locating hidden land mines, underground tunnels and weapon caches in Viet Nam; finding missing persons and murder victims; locating lost pets; pinpointing buried foundations, walls and tunnels of ancient historical structures; locating buried treasure and ancient artifacts; and even finding the best fishing spots. These applications are well-documented in the literature (Bird, op. cit.).

Map Dowsing: Map dowsing involves working with maps to locate water wells, mineral deposits, oil fields or ancient archeological sites. This approach has advantages in that it eliminates the necessity of being physically present at the target site. In this case the dowser usually moves a pendulum over a map of the area in question to determine the most favorable locations for the specific objective. Map dowsing can save time and money, and can be accomplished thousands of miles away from the target location. Map dowsing is commonly used to pinpoint the most favorable sites and is usually followed up by dowsing at the physical location. Dowsing with pendulum or L-rods is also often practiced from the inside of cars, helicopters or airplanes. This approach allows the dowser to cover more ground than on foot over rugged terrain, and confirm map-dowsed locations before working at the targeted physical site.

Final Thoughts on Supersensonics and Dowsing

Within the context of Future-Science Technology and Supersensonics, it is my suggestion that dowsing be taught to children as part of their early educational curriculum. This process gives children a simple way to connect their intuitive fields to the quantum information field, and seek answers to questions. This helps to reinforce their natural psychic abilities by providing instant physical feedback. One of the wonderful things about this intuitive science is that anyone, anywhere in the world, can develop this mode of extrasensory perception, as it costs next to nothing to fashion a pair of L-rods from stiff wire or rod stock.

Dowsing can thus be considered an effective tool for tapping into the quantum information field, as it merges the personalized parameters of the individual with the quantum field to receive effective answers. Working within the dowsing framework only requires quieting and focusing the mind to form specific questions of *who*, *what*, *when*, *where* and *how*. For those interested in learning more about the art and science of dowsing, the website for the American Society of Dowsers in Danville, Vermont can be found at: www.dowsers.org

MASTER KEY 2
Human and Planetary Eco-Evolution

"For global society to achieve enlightened transitions into the third millennium,
we must set aside our egocentric ways of living and thinking,
and focus on the interactions between humans and the natural world.
Only then will we begin to comprehend the process,
that culminated in the creation of human consciousness,
and properly recognize, honor and preserve,
the wondrous diversity and elegance of the intricate web of life on Earth."

Elliott Maynard, 2004

The Gaia Hypothesis, Global Ecology and Future-Science Technology

The Gaia Hypothesis was conceived in the 1960's by British chemist/inventor, Dr. James Lovelock. Lovelock envisioned a system of *global homeostasis* that was apparently sustained within specific life-range equilibrium by the harmonious interactions of the various components of the global ecosystem. From this perspective, the entire range of life forms – from viruses to whales – interacts with the atmosphere, landmasses and oceans to regulate Earth's biosphere, such that it remains at a level, which is optimal for the sustenance and continuation of life. The Gaia Hypothesis thus emerged, with the name *Gaia* being derived from the Greek name for the Earth goddess. This hypothesis suggests that the most satisfactory approach to gaining an understanding of our vast and complex planetary ecosystem is to regard the entire planet as one vast living ecosystem (*Gaia*).

In December of 1968, the astronauts of the Apollo 8 Mission to the moon experienced an event, which changed the course of human history. For the first time these privileged individuals witnessed a sight unlike any seen by humans before them – the enormous blue-green sphere of Earth, turning majestically against the blackness of space. From this moment on, man's perspective of his relationship to his home planet began shifting from a national focus to a new global consciousness. The photographs taken by Apollo 8 astronauts produced an image which, when transmitted through the global telecommunications network, has become a unique symbol for our times. This archetypal image [*Earthrise, or Blue Marble*] produced a fresh new whole-Earth symbol, which implied that mankind and his planet are part of a single system that supersedes individual, corporate or national interests.

Environmental Generational Amnesia

According to University of Washington psychology professor Peter

Kahn, children have a basic need to experience intensive interactions with Nature to enrich their physical and psychological well-being. Most adults already understand that our natural world suffers from environmental pressures. This is also true of our children, who find themselves growing up in increasingly bleak environments that are often far removed from the natural landscapes from which our ancestors evolved. According to Kahn, "Even more startling is the fact that we, as adults, hardly know this is happening."

Kahn believes that the presently impoverished state of our natural environment is partly due to a phenomenon he terms, "Environmental Generational Amnesia." This concept implies that most of us tend to accept the natural environment we experienced as a child as the *norm*, against which we tend to measure the environmental degradation experienced during our lifetime. Thus, with each successive generation – even though the amount of environmental degradation increases – the next generation accepts this degraded condition as their reference point of what is *natural*. According to Kahn, "The upside of this is that children start fresh, unencumbered mentally by the environmental misdeeds of previous generations. But the downside is enormous; in that children think what they encounter is the environmental norm. At some time they understand that the baseline is wrong, but they don't understand it at a visceral level."

The concept of Environmental Generational Amnesia emerged from Kahn's research, which studied basic environmental values and concepts of black children in Houston, Texas. At the time, Houston was one of the most polluted cities in America. Although Kahn remembers waking up in the morning, feeling "overpowered by the smell of oil refineries," the children he interviewed in the study often said there was no pollution in Houston – even though they were experiencing the same smell that Kuhn regarded as "overpowering."

Kahn also conducted studies with groups of children and young adults in Brazil, some of whom lived in urban areas, and other groups who lived in the Amazon jungle. He included data from another group of children in Lisbon, Portugal. In the locations he studied, he discovered that similar belief patterns existed in the children with regard to the natural environment. In other words, despite a deep ingrained need for healthy experiences and interactions with the natural world, people's experiences with diverse natural ecosystems are, as a whole, declining rapidly. Says Kahn, "We love Nature, need Nature, and are drawn to the natural world. Our connection to the natural world is so deep that some people drive for hours just to walk on the beach." Children everywhere require a diverse ecosystem and a variety of natural interactive experiences for their happiness and well-being. Although humans seem to have been reasonably successful in adapting to urban systems and impoverished conditions, this adaptation has apparently come at a tremendous cost. Thus, for the health and well-being of present and future generations, we must, as a society, make more positive and enlightened choices with regard to preserving the natural environment in its original state (Schwartz, 2003).

The critical nature of our basic human need to bond with nature is expressed in the following quote: "We all depend on a healthy planet to sustain life. There is something each of us can do to stop living in a world of more and start living in a world of enough. The individual choices that we all collectively make – to live within the means of our only planet – have the potential to create a better world, inspire our children and their children, and send signals to the world's leaders that this is what we desire" (World Wildlife Fund, 2013).

The Healing Effects of Forests and the Natural World

According to Dr. Eva Karjalainen of the Finnish Forest Institute, "Many people feel relaxed and good when they are out in Nature. But not many of us know that there is also scientific evidence about the healing effects of Nature." Dr. Karjalainen contends that forests and other natural settings can reduce stress, anger and aggressiveness, and increase a person's sense of happiness and well-being. Being out in the natural world has also been shown to strengthen our immune systems by increasing immune system activity and the numbers of natural cancer killer cells.

Studies have shown that people from natural settings recover from stressful situations faster than those from urban environments. Blood pressure, heart rate, muscle tension and levels of stress hormones have all been shown to decrease faster in natural settings. In children, symptoms of depression, anger, aggressiveness and ADHD have all been shown to be reduced when they play in natural places. In an interesting footnote to the environmental benefits on mental and emotional well-being, more than half of the most commonly prescribed drugs for these conditions are derived from tree bark. Notable examples include Taxol, a drug used to combat ovarian and breast cancer; and Xylitol, a sweetener used for combating dental decay.

With regard to the overall health-promoting benefits of natural parks and lands, Dr. Karjalainen says, "Preserving green areas and trees in cities is very important to help people recover from stress, maintain health and cure diseases. There is also monetary value in improving people's working ability and reducing health care costs" (www.phys.org News, 2010).

Man's Oldest Ancestor More than Twice as Old as Previous Estimates

New DNA evidence has recently revealed that our oldest known human ancestor may well be twice as old as formerly thought. This startling revelation, recently reported in *New Scientist,* was based on DNA evidence from a deceased African American man named Albert Perry who lived in North Carolina. Following Perry's death, a family member submitted a sample of his DNA to a Houston, Texas genetic testing company, with the objective of learning more about the family's lineage. Results of DNA analysis proved to have implications, which may well push back the date of human evolution to as much as 350,000 years!

The findings of this remarkable study, initially published in the *American Journal of Human Genetics*, showed that all previous studies of DNA samples had pointed to a Y chromosome, which traced back to between 60,000 and 140,000 years ago. The Perry DNA sample broke this trend by not matching up with the common ancestral line.

Jon Wilkins of the Ronin Institute in Montclair, New Jersey commented on this historically remarkable discovery: "It's a cool discovery. We geneticists have been looking at Y chromosome about as long as we've been looking at anything. Changing where the root of the Y chromosome tree is at this point is extremely surprising."

Following the initial series of tests on Mr. Perry's DNA, geneticists at the University of Arizona conducted further tests, and discovered that the Y chromosome from these tests matched up with 11 men from a single village in Cameroon. University researcher Michael Hammer stated that evidence suggests that Perry's DNA may have originated from an earlier species of humans that went extinct in the course of human evolution, but not before interbreeding with the more modern version of what we regard as modern man (Pfeiffer 2013).

Greenland Ice Sheet Melting

Imaging data for Greenland in 2012 brought to light some startling facts with regard to the melting rate of this massive ice sheet. Computer analysis of the data indicated that, from July 8 to 12, melting had accelerated dramatically, with an estimated 97 percent of the ice sheet showing surface melting. Data was collected by the three different satellites, which measured different physical properties on different scales, while passing over the same area at different times.

Although researchers were surprised at the amount of ice melt, which extended from the ice sheet's coastal edges to its 2-mile thick center, they were cautions about linking this accelerated melting of the surface ice directly to global warming. In conjunction with the satellite data, ice core samples from the highest point of the ice sheet indicate that the last time such a period of accelerated melting occurred was in 1889 (BBC News, 2012a).

∞

Earth's Glaciers Are Out of Balance

According to the latest research Earth's glaciers are out of balance. Scientists based this statement on a study of 144 large and small glaciers in different parts of the world. According to Sebastian Mernild, member of the Climate, Ocean and Sea Modeling Group (COSIM) at Los Alamos National Laboratory, "When we look at the data, we can see that the glaciers are out of balance, meaning that the climate has actually changed faster than the changes we've seen in ice area and volume." Furthermore, "Our data suggests that the glaciers will commit about 30% of their area, and about 38% of their volume to global sea level rise." In essence, this means that glaciers will tend to become smaller, as they move back up onto the land masses, and the melting ice will contribute to sea level rise (Amos, 2012).

∞

Global Weather Engineering: The History of Rainmaking

The basic technologies for weather engineering were developed over 50 years ago. In the late 1940's, mathematician John Von Newmann researched weather modification for the U.S. Department of Defense regarding potential applications in "climate warfare." In the 1950's Wilhelm Reich broke new ground in weather modification research with his "cloudbusters" in the State of Maine. In the 1960's Dr. Bernard Vonnegut worked to improve existing cloud-seeding techniques that used silver iodide crystals to bring rain to parched areas of the world.

Less known, is the research of Trevor James Constable, whose life work involved developing and refining what he termed "Etheric Weather Engineering." Using his position as a radio electronics engineer in the U.S. Merchant Marine he completed over 300 crossings of the North Pacific Ocean. During this period he conducted a series of weather manipulation experiments, using a specially modified multi-tubed cloudbuster – a refinement of Reich's original design. Concurrently, Constable developed what he called his "Etheric Vortex Generator," with which he demonstrated that local weather conditions could indeed be modified. This work was carefully documented via time-lapse video.

In the 1990's Constable took his weather engineering airborne, using small aircraft in which he installed his instruments. By 1994, in a series of aerial tests over Hawaii, he conclusively proved that airborne etheric weather engineering was a viable weather modification technology, since the wide areas of visibility and infinite heading controls of aircraft tended to magnify the effects produced earlier from his ship-based experiments.

By the late 1980's and 90's Trevor went on to demonstrate a series of smog abatement experiments in Southern California, which, at that time, was considered to have some of the worst air pollution in North America. In 1990, funded by Singapore entrepreneur, George Wuu, he established 14 ground stations for his "vortex generators" in Southern California. Surprisingly, smog levels for the full season were lowered by 24 percent below "normal" – at a modest cost of $35,000. Despite these positive results, the State of California decided it was not interested in pursuing this research further. This, despite the fact that the state subsequently spent some ten billion dollars by 2003 with minimal results.

In 1991, Constable and Wuu formed a weather-engineering corporation in Singapore called Etheric Rain Engineering, Pte. Ltd. Subsequently, they received a contract with the State of Melaka in Malaysia to help bring rain to this driest state. Atop a hotel in Melaka they claim to have engineered 38 measurable rains during a 57-day period during the dry season. This amount represented approximately 75 percent of a normal rainfall in the area for a full year (Adachi, 2003). The companies website is: www.rainengineering.com

Constable's research is documented in a 1994 book called *The Loom of the Future* (Brown, 1994). Among his numerous publications, Constable is most noted for his book, *The Cosmic Pulse of Life: The Revolutionary Biological Power behind UFOs* (Constable, 2008).

∞

What in the World Are They Spraying? The Chemtrail Conspiracy

The entire business of "chemtrails" is intriguing, convoluted, covert and complex. Queries to trusted scientific colleagues yielded little information, as, at the time of this writing, the subject is apparently the sort of "hot potato" item that is not only *not* respectable for mainstream scientists to study, but there is little incentive or funding for such research. Accordingly, I decided to include a brief section on chemtrails and present some of the information that *is* available, as the overreaching effects of spraying materials containing Barium and Aluminum cannot fail to directly impact our atmospheric and terrestrial commons, the health of vegetation, food crops and farmed animals, as well as our own health and quality of life. Ultimately, this type of megaproject involves massive amounts of money and equipment, and no doubt some interesting liaisons between governments, military forces, large corporate interests and researchers on the HAARP Project.

From my early years in Maine, I learned how to "read" the weather, as it was always an important factor in planning outdoor adventures, or predicting winter driving conditions. What we do know about chemtrails is this: Anyone who has looked up in the sky during the past couple of years will have seen networks of crisscross patterns which remain in the atmosphere for hours and coalesce to form cloud cover in areas where this would not normally occur. No, these are not ordinary "contrails" that are normally associated with commercial and military aircraft, as these contrails normally disappear a few plane lengths behind the aircraft, but under certain atmospheric conditions do tend to linger and form larger clouds. There have been some studies on the effects of contrails on the weather, but when we are talking about chemtrails, we are speaking about a phenomenon, which has occurred, on a large scale and over the past few years.

Rather than to open a Pandora's Box of complexities associated with the chemtrail phenomenon, I would suggest the excellent DVD's, *What in the World Are They Spraying?* and *Why in the World Are They Spraying*? These video presentations offer a comprehensive overview the chemtrail dilemma, as well as visual evidence of chemtrail-generated weather, soil and water contamination from sprayed metal compounds and aircraft using their spraying arrays (Griffin, Murphy and Wittenburger, 2010; Murphy, and Kolsky, 2012). For readers wishing to pursue the subject of environmental weather warfare, the following references should be useful: Weather Modification, 2014, Weather Warfare, 2014 and Weather Weapons, 2013.

Wind Turbines Can Modify the Weather, but Do Long-Term Benefits Outweigh the Environmental Consequences?

There are most certainly obvious environmental advantages for generating electrical power with wind turbines. Recently with the global proliferation of wind power arrays, several downsides to wind power have emerged. If these issues are not addressed, and if possible corrected, this new technology could produce a host of negative environmental factors with regard to weather, crop-health, damage to humans and other life forms, and a host of regulatory, legal and insurance repercussions.

Recent scientific studies have, for the first time, shown that wind farms *can* cause climate changes in areas near large turbine arrays. Normally, air close to the ground becomes cooler at night when the sun goes down and ground temperatures cool. On large wind farms, however, the turbines tend to mix the air so the overall effect is to raise the average temperatures in these areas. Satellite data from a large area of Texas (home to four of the world's largest wind farms) has indicated that for over a decade, local temperatures were raised by nearly 1°C as more and more wind turbines were installed. The global implications are that these wind power arrays could affect regional weather patterns. With Texas presently the largest U.S. wind power producer, and with China installing an average 36 wind turbines each day, problems of weather disturbance from wind turbines could become an environmental problem which needs to be addressed and remedied by modifying the existing technology.

According to Liming Zhou, Professor at the Department of Atmospheric and Environmental Sciences at the University of New York, "Wind energy is among the world's fastest growing sources of energy. The U.S. wind industry has experienced a remarkably rapid expansion of capacity in recent years. While converting kinetic wind energy into electricity, wind turbines modify surface-atmospheric exchanges and transfer of energy, momentum, mass and moisture within the atmosphere. These changes, if spatially large enough, might have noticeable impacts on local to regional weather and climate."

Negative Impacts of Wind Turbines and the "Wind Turbine Syndrome"

In a survey conducted in a Waterloo, South Australia wind farm, as part of his Master's Degree research, Zhenhua Szn determined that 70% of residents living up to five kilometers away from wind farms reported being negatively affected by wind turbine noise, with over 50% of them being "very or moderately negatively affected." This data suggests that the effects of these specific wind turbines are considerably higher than previous studies in Europe had indicated. In one example, a neighbor, Andreas Marciniak, who lived near the Waterloo Wind Farm expressed his feelings in a letter to a local newspaper as follows: "Do you think it's funny that at my age I had to move to Adelaide into my mother's shed and my brother had to move to Hamilton into a caravan [trailer] with no water or electricity?" Marciniak and his brother were both advised by their doctors (including a cardiologist) to leave their homes and not return when the wind turbines are turning.

In Denmark, the government has taken the "wind turbine syndrome" seriously enough to focus its attention on low-frequency noise (LFN), and impose regulations to limit LFN from wind turbines to levels that are tolerable to humans and animals. The symptoms include: insomnia, headaches, nausea, stress, inability to concentrate and irritability, which led to declining health, vitality and reduced immunity to illness. The wind turbine issue has been subject to heated debates, since residents have been forced out of their family homes and off their land, simply because it has been difficult or impossible for them to function in the vicinity of wind turbines. This has become enough of a public issue that two organizations have been formed to give the public a voice against the negative effects of wind turbines: The European Platform Against Wind Farms (EPAW) and the North American Platform Against Windpower (NAPAW). In the words of NAPAW spokesperson Sherri Lange, "It'll take time to gather enough money for a big lawsuit, but time is on our side: victim numbers are increasing steadily" (Duchamp, 2012). The EPAW and NAPAW websites can be found at: www.epaw.org/index.php?lang=en&page=4 and www.na-paw.org

In conclusion, as worldwide proliferation of wind turbine arrays continues to expand exponentially, there are still many unanswered questions, which remain. There are other aspects such as bird kills, which need to be considered and resolved. The global weather system is an incredibly complex system, and when the new wind turbine arrays are put into operation, this has the potential to disrupt weather patterns and thus change such aspects as prevailing winds, precipitation and other meteorological factors. Wind power proponents continue to downplay any negative effects, but their main argument is that these negative impacts are minor when compared with the positive environmental impacts of wind power as opposed to the polluting effects of coal and oil-fired industrial and electrical power generating facilities. As windpower technology continues to proliferate, more data on its effects on humans, plants and wildlife will continue to surface. It should then be possible to find new ways to reduce or eliminate the present negative impacts of wind turbines on human health and birds.

The Oldest Trees on Earth

One of the world's oldest trees is a Bristlecone Pine Tree located in the White Mountains of California. This tree has been carbon dated at 4,841 years, and until very recently was thought to be the world's oldest non-clonal organism on our planet, although another Bristlecone Pine dated at over 5,000 years old was mistakenly cut down by a researcher (Wikipedia, 2013).

In April of 2008, another fascinating discovery was made. On a mountaintop in Dalarna, Sweden, researchers found four generations of spruce cones, which, when dated at a Florida lab, came out to be 375:5,669:9,000 and 9,550 years old. In other words, the oldest trees began their life cycle at the end of the last Ice Age (*Science Daily*, 2008).

Scientists Revive a Flowering Plant after 30,000 Years

Recently, Russian scientists were successful in propagating seeds found in a squirrel's burrow, buried beneath 125 feet of permafrost in northeastern Siberia. The resulting white-flowered plant, *Silene stenophylla*, was grown under laboratory conditions using seed and leaf tissues which had been stored away thousands of years ago and subjected to a quick freeze. Using a process called clonal micropropagation; researchers were able to produce 36 healthy plants in their lab. This is remarkable, considering they date back to the Pleistocene Age – a time before agriculture, and even the end of the last Ice Age. When the plants were radiocarbon dated, their age was determined to be 38,000 years. Results of the study were published in the Proceedings of the National Academy of Sciences.

According to researcher Svetlana Yashina at the Russian Academy of Sciences, "We consider it essential to continue permafrost studies in search of an ancient genetic pool, that of pre-existing life, which hypothetically has long-since vanished from the earth's surface." The scientists speculate that permafrost, which makes up about 20 percent of the earth's surface, could represent a time capsule – a vast resource where ancient life forms are preserved. If these life forms can be revived they would yield a wealth of information about the history and evolution of life on earth. The scientists also praised the Svalbard Global Seed Vault in northern Norway, a high-security facility, constructed as a low-temperature repository for seeds of all edible plants. To date, this repository contains over two million seeds. The facility represents a sort of frozen Noah's Ark, where over 100 nations have already contributed seeds, so these species would be available in case of some major global catastrophe (Potter, 2008).

Extreme Life-Forms Redefine Our Definition of "Life" and Offer New Implications for Astrobiologists

During the past decades, researchers have begun searching for life in unlikely places where extreme environments exist. Such extreme environments include: Deep ocean depths, deep-sea smoker vents, extreme cavern environments, toxic mining dumps, highly radioactive niches, deep drilling samples and samples from ancient Antarctic lakes.

Recently, scientists have made an amazing set of discoveries in extreme environments, which has expanded the limits for life at extreme temperatures, pressures and darkness. In February of 2013, bacteria were discovered a half-mile below the Antarctic ice. According to Anna-Louise Reysenbach, microbiologist from Portland State University's Center for Life in Extreme Environments, "Nobody realized there is so much biodiversity." She added that life in unexpected places "helps inform us what is possible elsewhere in the universe" (Rockoff, 2013). Further discoveries were presented later that same year when water samples from Antarctica's glacial lake, Whillans, which lies beneath some 2,625 feet of ice, yielded over a dozen species of "chemoautotrophs." These extremophiles use carbon dioxide, sulfur and ammonia as energy sources (Oskin, 2013).

Other examples include the following: In 2010 scientists exploring a cave in Chile's Atacama Desert discovered an algae, *Dunaliella algae*, which can exist with very little water. This new species survives in deep caves by absorbing dew that collects on the surfaces of spider webs. Other bizarre life forms include a group of hyperthermophiles that somehow manage to thrive in extremely hot environments. One such species includes bacteria belonging to the genus *Aquifex*, found in Yellowstone National Park hot springs, where temperatures reach 205 degrees F (96 degrees C).

Ocean thermal-vent ecosystems include barnacles, delicate shrimp, crabs, mussels and tube worms. These bizarre creatures exist in low-oxygen environments and have developed entirely new metabolic pathways, which are not dependent on light for photosynthesis.

Another class of extremophiles includes salt-tolerant microorganisms that exist in high salt concentrations that would kill conventional life forms. One example includes the bacterium, *Halobacterium halobium*, discovered in Owens Lake, California. This microbe has evolved to live in saline environments, which contain ten times the salt content of normal seawater.

Microbes called psychrophiles have been discovered in samples from the polar ice caps, glaciers and deep-ocean waters. These organisms have evolved special enzymes to function at below-freezing temperatures, and include bacteria, fungi and algae, which can live at temperatures as low as 5 degrees F (-15 degrees C). Another group of extremophiles include *Endoliths*, a group of organisms adapted to live inside rocks and in the pores between grains of minerals. These organisms have been discovered up to two miles (4.4 km) below the surface of the Earth – an environment considered toxic for humans (Slade and Radman, 2011). Since water is a scarce commodity at these depths, research suggests these organisms have developed adaptive metabolic pathways to feed on iron, potassium and sulfur (Moscowitz, 2011).

One of the most fascinating groups of extremophiles includes bacteria that have evolved in naturally occurring radioactive sites. Apparently, the natural evolutionary process has been speeded up since the proliferation of nuclear weapons and reactor technology in the 1940's. Research over the past few decades have led scientists to believe that exposure to uranium and other radioactive materials can lead to the creation of entirely new species. For example, In South Carolina, at the 300 square-mile Savannah River Clean-Up Site, strange white cobwebs were discovered. In a subsequent report filed by the Defense Nuclear Facilities Safety Board concluded: "The growth, which resembles a spider web, has yet to be characterized, but may be biological in nature."(Moscowitz, op. cit.). In another example, scientists, in 1970, discovered several types of a formerly unknown black mold thriving in high-radiation areas or near the Chernobyl Nuclear Plant in the Ukraine.

The bacterium, *Deinococcus radiodurans*, is one of the most radiation resistant species of extremophiles. This naturally occurring species has been genetically engineered to help clean up nuclear waste products. *D. radiodurans* is perhaps best known for its unique ability to repair massive DNA damage. It has also shown high resistance to DNA-damaging agents such as ionizing and harmful UV radiation, desiccation and mitomycin C, which inflict oxidative damage to DNA and other cellular components. Interestingly, these findings challenge the concept of DNA as the primary target of radiation toxicity, and highlight protein damage and the protection of proteins against oxidative damage as a new approach for dealing with radiation damage to living systems. It is significant that this organism has the ability to resist and break down nuclear waste in environments that would be toxic to normal organisms (Kazan, 2009).

The upshot of all this is that we now know that the bizarre marine flora and fauna in the deep ocean trenches and smoker vents, polar extremes and toxic waste dumps of our planet hold the keys for revising our concepts of "life" here on Earth. The textbook-defying metabolic pathways of these organisms also open new possibilities for the existence of extraterrestrial life-forms in places such as Jupiter's moon, Europa and the South Pole of Mars.

Why Whales are a Key to Sustainable Ocean Fisheries

The recovery of global whale populations during the past few decades would also seem to be good news for coastal fisheries. New studies have shown that the feces of these ocean giants add significant amounts of critical fertilizers to the oceans. In the Gulf of Maine, for example, it was found that whale feces add some 23,000 metric tons of nitrogen to this area each year, which represents more than the nitrogen contributed by the rivers, which drain into this ocean basin. It is already well known that certain species of microbes, plankton and fish recycle nutrients, but only recently has it been revealed how whales bring nutrients back to the ocean surface from the depths where they feed.

Biologists Joe Roman and James McCarthy have published a paper, which describes how whales and other marine mammals have historically played a key role in the productivity of ocean ecosystems. This is important since many areas of the Northern Hemisphere suffer from a limited supply of nitrogen, since phytoplankton tends to use up all available nitrogen during the summer months. According to Roman, "We found that whales increase the basic productivity. That leads to more phytoplankton, which 'pushes up the secondary productivity' of the critters that relay on the plankton." The result: Bigger fisheries and higher abundances throughout regions where whales occur in high densities." Roman states that conservative estimates by the world scientific community would suggest that former historical whale populations have been reduced to about 25 percent of former historical levels. Through studies of existing whale genetics Roman and McCarthy estimate these numbers may actually be as low as ten percent. This in the face of recent climate shifts, which have driven up water temperatures, causing a further decline in those nutrients that produce phytoplankton blooms.

One major implication of this new study is that this information strengthens the case of marine scientists to counter whale harvesting by governments that still allow whale hunting. In the words of Roman and McCarthy, "For a long time, and still today, Japan and other countries have policies to justify the harvest of marine mammals. The main argument of these countries is that whales compete with their commercial fisheries." According to Roman, "Our study flips that idea on its head. Not only is that competition small or nonexistent, but also the whales present can actually increase nutrients and help fisheries and the health of systems wherever they are found. By restoring populations we have a chance to glimpse how amazingly productive these ecosystems were in the past" (Goodman, 2010).

MASTER KEY 3
Future-Science Education:
Creating New Learning Systems for Coping with the Present and Future Realities

> "Education should unfold full creativity and higher states of consciousness,
> to produce citizens capable of fulfilling their highest aspirations,
> while contributing maximally to the progress of society.
> By harnessing the nation's greatest resource – the unlimited creativity of its citizens –
> effective education can ensure national prosperity, international competitiveness,
> and a leadership role in the family of nations."
>
> *John Hagelin, 1998 – Manual for a Perfect Government*

In 1970 Alvin Toffler had this to say about conventional educational systems: "What passes for education today, even in our 'best' schools and colleges is a hopeless anachronism. Parents look to education to fit their children for a life in the future. Teachers warn that a lack of an education will cripple a child's chances in the world of tomorrow. Government ministries, churches, the mass media...all exhort young people to stay in school, insisting that now, as never before, one's future is almost wholly dependent upon education. Yet, for this rhetoric about the future, our schools face backward toward a dying system, rather than forward to the emerging new society. Their vast energies are applied to cranking out industrial men...people tooled for survival in a system that will be dead before they are. To help avert future shock, we must create a super-industrial education system. And to do this, we must search for our objectives and methods in the future, rather than in the past" (Toffler, 1970).

The Arcos Cielos University of the Future Project is a new paradigm for a life-relevant educational system. This new operating system provides a set of tools for integrating *the art of teaching* with *the art of learning* into all levels of education.

The University of the Future Concept combines sets of *opposites*, which generate unique synergistic effects when combined with conventional educational programs. Examples include: ancient and modern worldviews, traditional and alternative medicine, eastern and western philosophies, conventional and alternative technologies and physical and multidimensional realities. When two such opposites come together a third new energy field is formed which combines the best qualities of both opposites, and thus has the potential to create new solutions for problems common to both opposing mindsets. In ancient times, clashes between two different cultures or religious factions most often resulted in bloodshed and warfare. Once the initial violence died down, however, cross-cultural exchange tended to generate periods of cultural enrichment and the sharing of the best ideas embodied in both cultures. Thus, when opposing ideologies and thought-locked institutions can be nurtured from this perspective, the resulting creativity and productivity can birth powerful new ideas. These ideas incorporate the best elements from all parties concerned, and generate win-win practical solutions. This powerful synergy offers the capabilities for transforming Marshall McLuhan's global village into an enlightened, ecologically prosperous and energy-sustainable global super-nation.

In the context of a new-millennium consciousness, traditional ego-based differences between global factions can be set aside and replaced with efficient *round table meetings* of core leaders who are able to envision and facilitate realistic change for the benefit of all factions. This kind of enlightened approach to problem-solving can produce new systems and solutions for the human-created problems that now impact our global biosphere. Within the thinking strategy of Future-Science Technology is embedded a key concept – that every problem has a viable solution that simply needs to be unlocked and nurtured for it to manifest.

The University of the Future Concept

Traditional education focuses on past events, names and time frames. It is the product of an industrial-age mindset. The analytical, left brained thought processes of this system emphasize learning names, dates, facts and figures. Students are evaluated for their ability to parrot back these facts, as opposed to learning through experience and problem solving in real-world situations. Traditional education focuses on analyzing concepts by breaking them down into their component parts instead of studying them in their dynamic real-world contexts.

The University of the Future Project offers unique, educationally efficient and cost effective alternatives to existing educational systems. Sadly, economic and social pressures have co-opted many of our existing university systems. Many universities have become glorified education mills replete with country club amenities and bloated athletic programs – all at the expense of educational excellence. Unfortunately, their major focus has too often been on perpetuating educational bureaucracies overstuffed with academic deadwood. By contrast, the Future-Science Educational System focuses on creating learning environments based on excellence, meaningful productivity and each student's unique needs – the ultimate in personalized education, with a major emphasis on *learning how to learn*. Students are thus provided with a supportive learning community offering them a customized set of life-management skills, designed to increase their chances of success and survival in real-world society. In addition to a balanced foundation of basic knowledge and communications skills, learning programs are *career directed* according to each student's abilities, interests and aspirations.

As a contemporary educational community, the University of the Future Concept is *supportive* and *synergistic*. The system emphasizes outcome-based experiential learning, which is anchored into the creation of new projects and business ventures. Within this neo-renaissance learning community adventures in self-discovery and development of the human potential are encouraged through apprenticeship and teamwork relationships between students, instructors and parents. Programs encourage *cross-cultural hybridization* with the objective of generating new ideas and perspectives. *High-tech* is combined with *high touch* and intelligent management of financial resources takes a high priority. Other objectives include: development of maturity, acquisition of wisdom, practicing compassion and support for others and development of the intuitive senses. These disciplines are interwoven into a living fabric, which expands to encompass higher dimensions, and include the development of intuitive spiritual guidance.

The University of the Future Concept regards the past as an exemplary resource to be mined for wisdom and experience, as opposed to over-emphasis on the parroting back of historical facts. New learning programs are focused on present events and strategies, keeping in mind that every action in the present will impact the future of the human race and the global biosphere for the generations yet to come.

Communications Fluency and Conscious Evolution

In the course of its social evolution human culture passed a series of milestones, including development of a written language, which was initially restricted to priests and scribes. Mass media was ushered in with the invention of the movable type Gutenberg press, an event which empowered a major shift from right-brained intuitive thinking to a linear left-brained modality. The progression of communications devices which were subsequently developed include: the typewriter, teletype, telegraph, telephone, radio, television, fax machine, computers, digital cameras, camcorders, cell phones and tablets. With the advent of transatlantic cables, satellite technologies and wireless routers, the global internet was born. It emerged as a unique electronic manifestation of the global social consciousness – "the global brain."

Within the new internet/communications context the following concept can be integrated into the organizational matrix for the University of the Future: If a *picture* is worth a thousand words, a *video* (or movie) is worth a thousand pictures and a *direct experience* is worth a thousand videos. Accordingly, all students within the new Future-Science educational system would be supported to achieve a basic level of communications fluency in written and verbal communications skills, photography, videography, photo-manipulation, website design, app creation and usage, and presentation technologies. These basics could be supplemented by virtual reality, gaming technology, holography and multimedia presentations.

Cyberlearning

Cyberlearning integrates the technologies and resources of computer science, digital imaging, telecommunications and distance learning. The global internet and its associated technologies have endowed cyberlearning with the capability to become an educational resource of unprecedented power and game-changing potential.

As an interactive educational tool, cyberlearning is financially efficient, and adaptable to any working or parenting schedule. Since cyberlearning is self-paced, it can be pursued on either a full- or part-time basis. Cyberlearning programs are normally set up in conjunction with existing educational institutions, or drop-in centers where computers, live instructors and fellow students are available during normal working hours. Cyberlearning is thus a powerful new educational resource for home schooling, and also valuable tool for adult-focused lifelong learning programs.

Other aspects of cyberlearning include digital information management, which combines digital communications, the internet and written or audio-visual presentations. Digital information management requires a familiarity with these resources, but most importantly, the ability to learn effective data mining strategies for locating, prioritizing, storing and organizing information into appropriate formats. Other aspects of cyberlearning include a familiarity with productive uses of social media, e-commerce, basic website design and operation and explorations into virtual reality.

Virtual reality, combined with digital communications technology offers tremendous potential as a learning resource. Although virtual reality was first developed by the military for strategic combat training, VR environments have been expanded into the field of videogames for people of all ages. This technology offers exciting possibilities for immersive education – especially when combined with *natural reality* (field work in natural settings). Virtual reality simulators provide a substitute for piloting military aircraft, spacecraft, deep-submersible vehicles or heavy construction equipment. VR simulators provide cost-effective training by reducing the need for real-time operation of multimillion-dollar equipment.

As virtual simulators continue to evolve, their applications will become more common in medicine, environmental sciences and space technology. Already, in medicine, human-controlled robotic manipulators can surpass the capabilities of direct surgical contact between surgeons and patients. Telerobotic manipulators have become so precise, for example, that human heart bypass surgery using endoscopic techniques has become routine. Since only small openings are required for inserting the endoscopic tubes into the chest cavity, patients are able to return quickly to their normal life following a short hospital stay for post-surgical observation.

In a typical operating theater scenario, the surgeon sits near the patient at a robotic console which is equipped with hand and foot controls. The surgeon receives tactile feedback from the variable resistance controls and visual feedback from a video monitor, which can shift views via voice commands to provide real-time feedback for the procedure via tiny video cameras inserted through the endoscopic tubes. Although such procedures normally take place with the surgeon and patient in the same room, telecommunications technologies can also link up doctors and patients located in different parts of the world.

Video gaming technology also represents a powerful learning tool for training the reflexes and mental thought processes. Although initially designed for the younger generations, video games have recently emerged as a significant family and adult pastime. The variety of games runs the gamut from military tactical warfare to role-playing, building cities, governing kingdoms or terraforming planets. *Sims* offers infinite possibilities for creating and exploring virtual worlds, managing virtual families or raising *cyber-pets*. These digital pets have the capability to grow, learn by trial-and-error, reproduce and die. Although the transformation of videogaming into a learning tool is still in its early stages, the compression of an entire interactive virtual world into a single compact disk is undeniably one of the greatest miracles of our times.

The Global Internet and "Webucation"

The global internet has become a major transformative force for third-millennium society. This amazing technology effectively collapses time and space in the sense that it allows anyone to communicate, interact and transact business with others anywhere on the planet. The internet never sleeps. It functions tirelessly 24-hours a day, seven days a week. Information stored in cyberspace can be accessed at any time, from any location. As an equal-opportunity communications medium the worldwide web dissolves social, national, ethnic and economic barriers.

Evidence also suggests that the internet has increased energy efficiency and reduced atmospheric pollution. Apparently, e-commerce, e-mail, video-conferencing and distance learning have served to eliminate a significant percentage of physical travel, and have thus increased the functional efficiency of certain areas of corporate organizational structure (Roman et al., 1999).

Webucation implies the inherent capability of the global internet to function as a self-learning electronic entity. Through a process of "digital selection," the website has applied its inherent programming to evolve as a human/artificial intelligence interface. From a future-science technology perspective, a website is a self-aware communications interface between humans, computers and the global consciousness field. Websites have also gained the innate capability to adapt and evolve, according to the digital input and interchanges that occur within their electronic consciousness field. Indeed, the website would appear to be developing into a *cyberspace entity* in its own right. Considering the transformations which have already occurred, it is only a matter of time before the basic website concept shifts from a two- to a three-dimensional medium, which incorporates virtual reality and holography – eventually providing immersive interactive experiences which will come to rival the holodeck on the Starship *Enterprise*.

Today, across America and in countries all over the world, the latest computer, tablet and cell-phone technologies are being stuffed into backpacks and set up on classroom desks, as students' access and organize vast amounts of data via hi-speed wi-fi systems. Although internet technologies, which integrate electronic and social skills, have been in place for several years, the situation at the university levels is different. With recent rises in tuition fees, and difficulties associated with repayment of student loans, virtual universities are coming increasingly into favor. Some educational experts even go so far as to suggest that communications technology may even eliminate the need for many traditional campuses altogether.

Virtual Universities Gain New Ground

Among the innovations online education can offer is included training in technical remote viewing. All that is necessary is access to a computer and a hi-speed modem. A company called Psi Tech created a unique virtual university interface for teaching technical remote viewing (TRV). The virtual campus is set up to simulate a physical campus, where students can access the various facilities. TRV's virtual campus includes an administration center, training labs, library, lecture hall and auditorium. Students visiting the virtual campus for the first time will encounter a virtual guide called "WANDA" (Website and Navigation Assistant), who takes them on a campus tour. At the administration center new students are issued an ID card, which allows them to register, pay tuition and attend classes. There is even a cafeteria, where students can get a virtual drink and a bite to eat.

Psi Tech claims to have created a unique learning environment, within which state-of-the-art tools are available for accelerated learning. The new teaching methodology uses something called "forced compliance." Unlike traditional home study methods like learning from DVD's, students will be able to skip ahead, but will ultimately be expected to complete all the basic lessons. Testing takes place each step of the way. The goal of the new learning system is to install new skill sets as quickly and efficiently as possible. This way, a rigorous training structure is insured and students can become qualified as the finest remote viewers in the world.

Using this new educational system, graduates from TRV University will also be able to advance in the latest remote viewing techniques to become certified. With instant access to TRV instructors and tutors can resolve problems and address questions as they arise. Thus, anyone with a computer and internet connection can become proficient in technical remote viewing, and students have access to thousands of their TRV peers around the world (Psi Tech, 2014).

Creating a Synergistic Future-Science Learning Community

Creating synergistic learning communities involves designing environments that function as *conscious* learning ecosystems – educational communities with *heart* and *spirit*. Spiritual development and the creation of an enlightened group consciousness are key elements for building this type of conscious learning community. Two additional key aspects of these learning environments include a sacred regard for the global environment and respect for the personal space of others.

In addition to the more esoteric aspects of the University of the Future Program a basic set of mind/body developmental skills are included. These skills include: Conscious food preparation, proper eating habits, advanced nutritional education, sports and recreational skill development, yoga and the martial arts, basic maintenance and repair skills and Paraphysical Fitness, an approach to fitness, which integrates physical and mental training with the higher consciousness centers [covered later in this book].

Other aspects of this holistic education program, often completely absent from conventional educational programs, include a balanced set of life management skills which include, proper manners, appropriate social dress and demeanor, time management, financial management, career development, intelligent mate selection, family management, reproductive responsibility and parenting skills. These values would be grounded by a group-generated code of behavior, based on honesty, integrity, commitment and a karma-based value system, with *karma* being defined as the ultimate example of self-responsible behavior – "What goes around comes around." Christianity's Golden Rule or the Buddhist Eight-fold Noble Path can serve equally well as responsible behavioral prototypes.

Subtle-Energy Management: Key to Future Education

Another major component of the University of the Future Concept is the understanding and practical application of subtle-energies in all aspects of our lives. By achieving a basic knowledge of subtle-energies, anyone can learn how to effectively operate at the interface between physical and quantum-field realities. They can thus develop and expand their own innate capabilities that have already been programmed into their genetics. Considering the recent merging of alternative and conventional medical practices, this kind of intuitive approach is already being put into practice by healthcare professionals the world over. Similar approaches in the fields of science, technology and education are yielding correspondingly dramatic innovations in their respective areas.

Subtle energy phenomena incorporated into the University of the Future Concept include: *Earth Energies*, an understanding of natural earth energies and their effects on the human system, and *Supersensonics*, the awareness and integration of our natural intuitive sensory abilities as applied to business, science and our daily interactions with others. Other subtle-energy phenomena include *Environmental Energy Enhancement* – the focus and magnification of natural earth energies. This can be achieved by combining ancient sciences such as Feng Shui and Geomancy with modern technology to create enhanced, interactive centers for personal enlightenment, planetary healing and breakthrough thinking. The awareness and practice of subtle energy work is thus a basic prerequisite for expanded consciousness development. This awareness can be achieved through special exercises designed to increase mental focus, concentration and creative thinking. Martial arts yoga and related consciousness disciplines can also be integrated into educational curricula to develop the mental, physical and subtle-energy abilities and expand the intuitive powers.

Rewiring Your Brain to Learn a Language in 10 Days

Learning a new language allows us to better understand the worldviews of other cultures. Learning new languages increases our professional skills and expands our horizons when we travel to other countries. Dr. Paul

Pimsleur has developed a unique language learning method, which has been successfully used by the FBI and similar government agencies. He discovered that in most countries individuals use only about 2,500 simple words and phrases on a daily basis. Using this as a basis for a new compressed language-learning program, Dr. Pimsleur created The Pimsleur Approach, which focuses on learning these 2,500 words and phrases in the shortest time possible. The system uses audio recordings and claims to be what language learning should be – quick, fun and easy. Interestingly, the course requires only a CD player, and does not involve any reading, writing or computer usage. This unique method for quickly learning the basic languages is claimed to have been used by over 25 million individuals, and even has a 100 percent money-back guarantee if the customer is unable to speak the target language in ten days (Pimsleur, 2012).

Conclusions and Final Thoughts

When the first television pictures of Earth were beamed back from space, this event represented a major step in the self-awareness of our planet – the point at which the *global brain* achieved self-consciousness. Considering the rapid development of communications technology since that time, the nations and communities of our planet have become electronically transformed into Marshall McLuhan's 1960's concept of a "global village." The rapid development of the telegraph, telephone, radio, television, fax, global internet and satellite networks have together created what might best be called a "self-aware global intelligence" (The Global Brain). The one-room schoolhouse – historically the educational icon of rural America – has metamorphosed into something inconceivable only a century ago. The computer has been transformed into an educational portal to the global brain – "a global schoolhouse-in-a-box!"

The University of the Future Concept thus embodies the development of high-energy, intellectually stimulating and emotionally supportive learning environments, designed to teach and nurture students and teachers alike. Within immersive experiential environments students undergo learning adventures, which emphasize creativity, self-responsibility, maturity, spiritual development, conscious learning and the pursuit of happiness. For a paradigm to be dynamic and timeless, it needs to be flexible, evolvable and grounded into practical reality. These objectives are fundamental to the University of the Future as an educational program suitable to meet the real-world educational challenges for the third millennium.

MASTER KEY 4
Future-Science Art:
A Quantum Technology for Creating "Living Artworks"

> Fear not the strangeness you feel.
> The future must enter you,
> long before it happens.
> Just wait for the birth,
> for the hour of new clarity."
>
> *Rainer Maria Rilke*

Future-Science Art is a transformative new paradigm for creative artistic expression. This concept integrates ancient and modern worldviews and highlights the blending and manipulation of subtle energies. The concept integrates linear left-hemispherical thinking with intuitive right-hemispherical thinking, thus opening the doorway for inspiration from the higher realms. This process reveals how *any* artist can develop a working relationship with the universal quantum field. Essentially, Future-Science Art is a method for creating a new type of artwork that integrates human consciousness and the quantum field.

By acquiring a heightened awareness of subtle-energy fields, anyone can learn to become a conscious co-creator of living artworks. Since each artistic creation embodies its own distinct set of *energy coordinates* these living artworks can take on a consciousness of their own. The final arrangement of the basic elements for each artwork, when assembled into a monolithic whole, creates a "multidimensional energy matrix." This energetic matrix can thus become "an aware intelligence." By blending energies from the higher realms with the physical components of an artwork, artists are rewarded for their efforts by a corresponding expansion of their own consciousness.

Future-Science Art can incorporate energies from the ancient past and distant future. It integrates ancient and modern technologies to produce art that generates its own *consciousness field*. Via the internet, digital images and videos of these artworks can be transmitted, stored, shared and enjoyed by anyone anywhere on the planet. Media networks serve as *consciousness amplifiers*, which can assist in planting seed concepts in the hearts and minds of others.

The Historical Background of Future-Science Art

The great Renaissance art masters were gifted with super-normal abilities to translate beliefs, images, emotions or energy patterns from higher sources into artistic creations which stand alone in their beauty and inspiration. Masterpieces like Da Vinci's *Mona Lisa*, Michelangelo's Sistine Chapel frescoes or his inspiring marble sculpture, *David*, all embody this higher consciousness. When channeled through the minds and hands of these gifted artists this new consciousness served to quicken the consciousness of those who came into the presence of these masterworks. A unique quality of *aliveness* is still powerfully apparent in many statues of the Greek gods, Christian saints and oriental Buddhas.

Throughout the ages such art masterpieces have served as focal points for personal and religious inspiration. The amplification and broadcasting of their unique energies create an atmosphere of peace, comfort and healing within the temples, churches or government centers where they were located. Modern examples of this phenomenon can be experienced with the Washington Monument, Lincoln Memorial, Statue of Liberty or the presidential faces at Mount Rushmore. Each of these epic monuments embodies a unique sense of reverence, inspiration and national pride.

Sacred Geometry is the architectural expression of art. It is based on sacred mathematical proportions such as the Golden Mean. By applying sacred principles pyramids, obelisks or domed structures can be designed and strategically positioned to amplify natural earth energies via the principle of *geoharmonic resonance*. By placing these structures on specific earth energy lines (ley lines) or at their intersections (power centers), buildings, communities and entire nations can be vibrationally *tuned* to create coherent high-energy fields which function to stimulate fine arts, science, education and commerce. Since art and architecture transcend national, religious and cultural barriers, this type of consciousness-field constitutes a powerful force, which has the potential for establishing global peace, and facilitating cultural exchange and cooperation. Within this type of neo-renaissance environment a sense of timelessness exists, since normal space-time coordinates are altered. Artistic and scientific concepts from the future can thus be more easily manifested into the present reality. Future-Science Art is a concept that reaches over two thousand years into the future. It is from this quantum consciousness that Future-Science Art and the other key paradigms of Future-Science Technology were created.

Techniques for Creating Future-Science Sculptures

This group of techniques employs methods that involve the combustion of different gases under pressure, or high-voltage electricity, which provides heat for cutting or welding the metal components into a single integrated unit. Future-Science Art adds an additional dimension since the artist sustains a constant *awareness* of the dynamic energies, which come into play during the process.

The basic technologies include Oxy-Acetylene welding and brazing, and the bending and shaping of the metal components. Plasma or gasoline/oxygen torches are used to cut the steel components. The sculpture is finished by smoothing the welding joints, then applying paint or lacquer to stabilize the external surfaces.

Components Used in Creating Future-Science Sculptures

Future-Science sculptures can be assembled from a variety of component parts, each of which may have a special meaning. When assembled in the final format the resulting artwork becomes transformed into a unique entity – something entirely different from the sum of its parts. The basic components have been grouped into the following categories: **Historical Coordinates** include pottery shards, antique iron or similar artifacts of historical significance. **Geographical Coordinates** include stones, shells, and minerals or souvenirs collected from places of sentimental significance. Such items incorporate a mix of unique energies from the locations where they were collected. **Geological Coordinates** by contrast, highlight mineral specimens for their vibrational qualities rather than location. Geological coordinates include crystals, gemstones, minerals and natural rocks. **Religious or Mystical Coordinates** include items from sacred sites. **Natural Coordinates** include items from the natural world. Examples include seashells, driftwood, dried seaweed, colored leaves and beach rocks. **Technological Coordinates** include man-made items. Examples include marbles, crystal spheres, gears, bolts and washers, cut nails and metal gears. These coordinates can be found as flea markets, yard sales and old dumps and landfills, where discarded machinery, broken toys, antiques and mechanical or electronic devices can be "discovered." **Electronic Coordinates** include low-voltage lights and mechanical systems. The most refined artworks in this category incorporate interactive sound and light experiences.

Consciousness Techniques for Creating Future-Science Artworks

Over the years I have developed the following approaches for the conscious co-creation of "living artworks:" This involves incorporating unique energies and applying them in what I like to call the "creative manifestation process."

Assembling the basic elements of the sculpture or artwork is the *design phase* of Future-Science Art. Designs for each artwork can be sketched by hand or on computer. Alternatively, the artist gathers the component parts for the art piece, and begins arranging them in different ways. During this creative process the set of unrelated sculptural elements come together to form a completed artwork under the artist's guidance. **Psychic Clearing** involves mentally "clearing" your tools, your workspace and the sculptural parts to insure they are free of negative energies that might interfere with your work. This is accomplished by silent or spoken affirmations, or smudging with sage. Over time this process will become automatic. **Psychic Shielding** involves shielding your workspace from negative psychic energies. This is done by visualizing and maintaining a protective field around you and your workspace at all times. **Grounding** yourself is important to create a working balance between the higher sources of inspiration and the physical reality. A state of "mindfulness" needs to be established with regard to what is happening around you in the ordinary world. **Calling in your muse** is another key element for artistic success. Concentrate in sending out a strong request and hold this thought in mind. Your muse will make its presence known to you. Simply open your mind to this and try it! **Connecting with your Spirit Guides and helpers from the higher realms** is also important. Each artist has their own unique set of spirit guides and helpers. They need to be acknowledged to activate this creative resource. Spirit guides allow you freedom of choice and will serve as facilitators between you and the higher creative forces. **Using meditation to generate ideas and provide guidance** is a process that can open up infinite pathways to entirely new concepts for artistic expression. When you set aside regular times for meditation and daydreaming, you will soon discover that ideas trickle down from higher consciousness into your mind. **Making postulates and affirmations** involves making statements of intent. To achieve specific objectives, write down a list of these items in the form of requests and statements. Writing these statements of intent serves to clarify your mind and gently hold them in your consciousness. **Using dowsing techniques** is a core aspect of the creative process. For example, dowsing can be used to determine how the elements of your finished artwork should be put together or what paint colors should be used in the finishing process. With this technique the artwork can be assembled piece by piece. Dowsing is especially useful for getting "yes" or "no" answers during the creative process. **Programming your mind** is easy. Think of your conscious awareness as a computer. This way, you can program yourself to bring in energies from higher dimensions, move into the past or future or receive inspiration via dreams and meditations.

Manifesting Future-Science Artworks: The Co-Creative Process

The following techniques are intended to guide Future-Science artists during the manifestation process: **Operating within the quantum field** involves establishing a "creative space" for your artwork to manifest. In time, you will enter that special state of consciousness where things happen fluidly and your artwork takes shape automatically. When this happens you are working within the quantum field – the "space of creation." As you become increasingly more aware, you will be able to shift into this super-creative mindset at will. **Assembling your artwork one piece at a time** involves selecting the basic elements and arranging them in different ways. Begin by joining two elements together to form a single whole. Then, select another piece that complements the first element. Continue on with this process and get into a rhythmic flow. Before you know it you will have a finished product. Be sure not to over-mentalize. Allow your automatic "idea processor" to work for you. As your awareness increases you will sense a force, which seems to encourage your artwork to be manifested. Simply open your mind to this possibility. **Visualizing your completed artwork** allows an idea or vision to appear in your mind during quiet moments, dreams or meditations. When this happens, you might wish to take some notes, as these ideas have a way of vanishing quickly unless you write down details to refresh your mind at a later time. You can transfer your notes to a sketchbook and develop them into full-blown sketches, which will later become finished artworks. [I have sometimes visualized paintings or sculptures in completely finished form, although in most cases I simply allowed the artworks to take shape, using the free-association or the stream-of-consciousness techniques mentioned previously]. Future-Science Art is full of surprises – you just never quite know just *what* is going to happen, or exactly *how* it will take place. This magical element of surprise is what this unique process is all about. **Working with the quantum field** involves noticing that certain *patterns* come into play during the creative process. Become aware of *when* this is happening and take note of patterns that produce the best results for you. Eventually, you will find that you develop your own tool kit of styles and techniques. **Achieving simplistic beauty** involves setting the mind aside. As a Tibetan master once said, "We are not naturally stupid; we go to school to get that way." Considering that our linear brain has been "educated" for years it's no surprise that it may be difficult to quiet this linear mind and allow the non-linear mind to function in its full creative capacity. With practice we can bypass this linear thinking modality and allow our intuitive senses to flow unencumbered.

Pablo Picasso once stated that the most difficult thing he had to do was to learn to "paint like a child." Pure artistic expression means creating without preconceptions. We simply allow the energies of creation to flow through our minds and hands to form masterpieces which are uniquely our own. By embracing simplicity and incorporating this principle in our art, we will almost always end up with a unique finished piece.

Understanding that your artforms are "ready to be manifested" implies that the art you are creating already exists in the higher realms. It simply needs to be brought into physical reality. Each unborn artwork tends to exert its own "birthing pressure." Be aware of this process and encourage these thoughtforms to manifest. This way, you become a "portal" for bringing these unmanifested pieces to life. **Adding your signature and blessing** is the final step. When your artwork is finished, sign and date it. Bless each one in your own special way. This frees your art to have a "life of its own." Remember the wooden puppet, Pinocchio, who was lovingly crafted by the clockmaker, Gepetto, who wished for a "real son." When the wishing-star fairy blessed Pinocchio, he became a "real boy." Blessing your artworks and consciously infusing them with unconditional love creates what I like to call "The Pinocchio Effect."

Additional Aspects of the Creative Process

Learning How to Learn involves observing the learning process as it occurs. When you reach the point where you can create and learn at the same time, your artworks will take shape almost magically – right before your eyes – as if someone else was creating them. In time, this merger between the creative and learning processes will infuse your technique and continue to evolve rapidly. **Sticking with the positives, eliminating the negatives** involves going with what works for you and eliminating those elements or processes that do not work. **Working on several artworks simultaneously** simply implies that some ideas and concepts require more time to "incubate."

Often, creative thoughts and inspirational flashes will sift down into your mind while it is occupied elsewhere. You may find that when you shift to another artwork, new ideas will often unexpectedly come through for the first one. **Cultivating the idea of "play"** suggests dropping into your studio periodically with no preconceived ideas. Begin simply moving pieces around in an abstract way. Since "Goofiness is next to Godliness," new ideas often surface when you allow yourself to simply "fool-around." You may be surprised at the magical ideas which surface during this time of play. **Staying grounded** is important. If you remain in a higher state of consciousness all the time, nothing will ever manifest for you! Periodically, ground yourself down into the present reality by taking short breaks outside in the natural environment. Take your shoes off and get your bare feet on the ground for a few minutes, have a cup of coffee or tea, a smoke or simply take care of a household chore; then go back to work. Little breaks like these provide fresh inspirations. **Shifting your viewpoint** involves setting the intention to open yourself to new sources of inspirational flow. This creates a conscious space for your ideas to take shape. **Daring to venture into the unknown** means not being afraid to "entertain the outrageous." Experiment with new styles, approaches and techniques. An artist's greatest reward is to produce artforms that delight. It is not really important what others think of your art, as you are entering new uncharted territories. Understand that the joy of creation has already manifested itself within your heart. This sense of joyful discovery will stimulate your creative energies to reach ever further into the unknown.

Final Thoughts on Future-Science Art

Anyone can learn to experience art as a healing, enlightening and self-directed process. Future-Science art can be considered a "catalytic agent" for the enlightenment, healing and evolution of human consciousness. This unique process transforms the artist and those who view these artworks.

Future-Science art embodies co-evolution and transformation. As a conscious co-creator the artist works in harmony with the subtle energies of the universe. Within this new and exciting paradigm lies a very special gift, since anyone in this mindset can create his or her own family of "art children." By their very nature these creations have the power to radiate their energetic signatures out to others through the process of "direct consciousness transfer."

The author's primary objective for this project was to create a critical mass of over 100 future-science artworks. It was envisioned that a new evolved type of creative consciousness could be established for artists everywhere. I also postulated that each piece that was gifted or sold would bring into manifestation two new artworks…either for the author or for other artists who wish to create their own living artworks. An in-depth treatment of Future-Science Art and photos of my artworks are included in my book, *Future-Science Art: A Unique Paradigm for Creating "Living Artworks"* (Maynard, 2010). Two of these sculptures were also included in Kim Carlsberg's book, *The Art of Close Encounters* (Carlsberg, 2010).

MASTER KEY 5
Future-Science Music

> "Music is a higher revelation than all wisdom and philosophy.
> Music is the electrical soil in which the spirit lives, thinks and invents."
>
> *Ludwig van Beethoven*

As a child growing up in Maine, I was fortunate to be blessed with parents who encouraged me in the musical arts. From an early age my life was enriched with classical music. I got to know many of the classical masterpieces by heart. Early favorites included Prokofiev's *Peter and the Wolf*, Rimsky-Korsakov's *Scheherazade* and Debussy's *Afternoon of a Faun*. Although we lived in rural Maine we thought nothing of driving 90 miles to attend a community concert series in Portland, which had a beautiful auditorium and the seventh largest organ in the world. This concert series attracted some of the world's finest musical performers. After each concert I would take my program backstage and have it autographed by the performing artist or conductor. I had programs autographed by violinists Jascha Heifitz, Zino Franscatti, conductor Herbert Von Krajan and pianist Arthur Rubenstein among others.

I began violin lessons during fifth grade of elementary school. By the time I attended high school I got more serious with the violin. I played in the orchestra and began to practice from two to four hours a day. My violin teacher declared that anyone who could play the violin should also be able to play the piano. Since we had my grandmother's parlor grand in our living room I practiced piano along with the violin and learned to identify random notes played by my teacher with my back turned. At the time I had aspirations of becoming a concert violinist, but soon discovered this kind of career was not for me. I went on to other adventures, but in recent years I purchased a beautiful Zeta electric violin with crystal pickups. I added a guitar effects processor and a Jimmy Hendrix wah-wah pedal. This allows me to produce unique musical sounds that exceed the limits of an acoustic violin.

Future-Science Music deals with the creation of "living music." As with Future-Science Art, Future-Science Music is *interdimensional*. Essentially, the artist agrees to open himself up to new sources of inspiration from the higher realms and thus learns to work within the fabric of the quantum field. Just as future-science artists make basic postulates to infuse life-energies into their artworks, future-science musicians bring through musical compositions from the higher realms by establishing an *interactive quantum field* such that an *interface* is created between the energies of the higher realms and the field of the musician or musicians. A similar creative field is formed between musicians and their listeners. This results in a synergistic musical field that is uplifting, healing and powerfully inspirational. The *conscious* co-creation of music within this new modality represents powerful new tool for achieving inner peace and enlightenment. This "enlightenment effect" can be extended by psychically energizing recordings and simulcasts, which are broadcast out into the planetary consciousness field via the global internet and satellite TV networks.

The great composer-musicians of the past, such as Beethoven, Mozart and Bach were uniquely gifted with the ability to bring through music from the higher realms in creating their timeless masterpieces. Most often this was done unconsciously, as admitting that their music was channeled from the higher realms would have no doubt generated resistance from religious authorities of the times. Future-Science music provides a new twist to this ability since it embodies the concept of *consciously* tapping into and accessing these higher realms. This process involves cultivating an awareness of *how* and *when* this interdimensional bridge between the physical and higher realms is operating, and then learning to operate within this creative "zone." One key characteristic of musicians who have cultivated this gift is that, unlike most musicians who simply focus linearly on technique, future-science musicians continually "reinvent themselves" by creating musical compositions that are new, inspiring and provocative. Modern examples of this type of musician would include Art Vangelis, who has produced epic movie scores and musical events, which have served to elevate the human consciousness to new levels. Examples of his works include *Antarctica* and *Chariots of Fire*. Another modern musician whose work stands apart from his musical contemporaries is Jean Michael Jarre, whose groundbreaking compositions, *Oxygene* and *Concert in Houston* serve to raise the consciousness of live concertgoers as well as listeners who listen to the recorded performances.

One musician who openly channels his music is keyboardist Robert Coxon, who provides live music for large gatherings in association with "Kryon," an extraterrestrial entity channeled by Lee Carroll, noteworthy for his United Nations-related channeling sessions. Information on Robert Coxon and his inspiring music can be found on his website at: www.robertcoxon.com/life.html

How Music Tones the Brain and Improves Learning

According to Northwestern University researchers Nina Kraus and B. Chandrasekaran, learning to play a musical instrument has many potential benefits that include a better understanding of language and improved learning skills. Recent studies suggest that new connections made between brain cells during musical training and practice can enhance related forms of communication such as speech, reading and foreign languages. This research suggests we should re-examine the role of music in shaping our development, and that educational institutions should include more music training into their curriculums. As with physical exercise and sports, music would appear to have a significant impact on "auditory fitness."

Since a musician's ear needs to be finely tuned to musical sounds, timing and quality, scientists have determined that musical training elicits changes in the brain's auditory regions. For example, auditory brain tracings from pianists have shown that activity increases in the auditory cortex – the region of the brain responsible for sound processing. It has also been shown that musicians have significantly larger brain volumes in motor and auditory areas than non-musicians.

The advantages of musical training also extend to the ability to understand speech, as studies demonstrate that children who have had musical training exhibit more neural activity in response to changes in pitch during speech than those without musical training. This suggests that the enhanced abilities of musicians to detect pitch changes may also help them improve their language skills, as they seem to be better able to bridge the gap between sound patterns and words. Musically trained children have been shown to have better vocabularies and reading skills than children without musical education. Musical training also seems to help children with learning disorders like dyslexia, since they tend to be overly sensitive to background noise. Musical training apparently allows them to concentrate better in classroom learning environments (Rettner, 2010).

Musical Duets Lock Brains and Rhythms

Scientists from the Max Planck Institute in Berlin have demonstrated that brain synchronization occurs between musicians playing a guitar duet. In an article published in *Frontiers in Neuroscience*, researcher Johanna Sanger and her team attached electrodes to musicians and recorded wave patterns in different regions of the brains of guitarists while they played different voices of the same duet. The objective of this experiment was to determine whether the musicians' brains would synchronize if the two guitarists were not playing exactly the same notes, but instead played different voices of the same musical composition. The researchers used 32 experienced guitar players in duet pairs. Players began at the same time, with one of the pair taking the lead and the other following. Each pair showed coordinated brainwave oscillations, even when playing different voices of the same duet. This type of synchronous brain activity is called *phase coherence*.

According to researcher Johanna Sanger, "When people coordinate their own actions, small networks between brain regions are formed. But we also observed similar network properties between the brains of individual players, especially when mutual coordination is very important; for example at the joint onset of a piece of music."

This research indicates that synchronization between different individuals happens in brain regions that are associated with social and musical interactions. Sanger goes on to say, "We think that different people's brain waves also synchronize when people mutually coordinate their actions in other ways, such as during sports, or when they communicate with one another." (Sanger, 2012).

This research data also explains how professional musicians, during jam sessions, are able to enter into this state of phase coherence and somehow anticipate the actions of others in the group. Group chanting, provides another examples of how two or more individuals can form a creative field in a state of phase coherence or mutual resonance.

The Amazing Powers of Music Revealed

The link between music and athletic performance is just one example of the new ways medical professionals and scientists are beginning to understand the power music has over our bodies and minds. According to sports psychologist Costas Karageorghis of the UK-based Brunel University, "Music is a great way to regulate mood both before and during physical activity. A lot of athletes use music as if it's a legal drug. They can use it as a stimulant or as a sedative. Generally speaking, loud upbeat music has a stimulating effect and slow music reduces arousal."

Increasing numbers of health professionals such as Linda Fisher of Loyola University Hospital in Illinois routinely play therapeutic music for their patients in hospitals, hospices and clinical facilities to induce healing effects along with their medical protocols. [From my own perspective, music therapists or live musicians can further enhance this healing effect]. Says Fisher, "The music I play is not necessarily familiar. It's healing music that puts the patient in a special place of peace as far as the music's rhythm, melodies and tonal qualities." Other studies undertaken at the Bryan Memorial Hospital in Nebraska and St. Mary's Hospital in Wisconsin showed that music significantly reduced heart rates, and lowered blood pressure and respiration rates of patients who had undergone surgical procedures.

Scientists have also discovered that music evokes memories. It has also been found to help ease labor pains, reduce the need for sedation during surgical procedures and assist in alleviating depression. With regard to the healing powers of music, the following basic principle has emerged: "Frequency plus Intent equals Healing." In this instance, healing refers to putting something back into proper *resonance*. *Intent* can thus be considered to be an essential ingredient for directing healing frequencies. Intent is also a key element in actualizing transformation, resolution and manifestation. To become effective in these modalities it is important that our intent must be pure, focused and unconditional for true healing or manifestation to occur.

During physical workouts in the gym, listening to music can lower our perception of activity by about ten percent. Music has also been found to have a profound influence on our mood, as it tends to elevate positive mood aspects such as excitement and happiness, while reducing negative mood aspects such as depression, stress and anger. Music can assist us in setting the proper pace during long athletic events such as marathons and also help eliminate nervous jitters before competition (Beaulieu, 1987; Lloyd, 2008).

Visualization Plus Vocalization Equals Manifestation

Visualizations and thought forms originate from the higher realms. According to conventional scientific research, some 80 percent of human brain function is associated with visualization (Bennett, 2014). By adding the power of sound to our visualizations we can more easily bring these visualizations down into the physical. Since sound is inherently multidimensional it bridges dimensions via *harmonics* and *resonance*. Conversely, sounds, which are audible to humans, also resonate harmonically in the higher dimensions. Sound is thus *multidimensional*.

Harmonics are geometric multiples created by the specific vibration of an object or thought-form. Although the A-string of a violin vibrates at 440 cycles per second this vibration is actually a composite of the fundamental note plus its overtones all vibrating together. When we develop our ability to hear harmonics new brainwave pathways are opened up. By learning to truly "hear" harmonics, and then to consciously create them, we can open ourselves to experience astounding new things. By "learning to listen" we can dramatically heighten our perception and expand our consciousness. Vocal harmonics also offer a way to create "bridges" to the higher planes. Listening to harmonically predominant sounds – especially by trained chanters – can indeed be an uplifting and transformative experience.

Our brains function both as transmitters and receivers of light-encoded information. The brain functions much like a radio in that it can transmit and receive different frequencies. Since *frequency sets* exist like radio stations, it is simply a matter of tuning our brain receivers to certain specific stations or channels. Thus, sound is perhaps the easiest way to re-tune our brains. This will in turn, transform the frequency structure of our entire vibrational field.

There are several different "sub-chakras" within the head and brain. The *Alta major* region near the occipital area at the back of the head is sometimes called the "channeling chakra" because it has been shown to be electronically active when channeling occurs.

Whenever music is played fields are created. When we become increasingly aware of these musical fields we allow them to positively influence our consciousness. In addition to the musical sounds themselves, it is important to realize that the *silence* between the sounds is when transformation often occurs. Adding *movement* to music also creates its own fields and assists in anchoring light encodements from the higher realms down into our physical and higher bodies. This is why dance, martial arts or Tai Chi Katas, performed to music are powerful tools for healing, grounding and tuning the physical body and higher consciousness centers.

Consciousness Attunement and DNA Activation via the Sacred Solfeggio Scale

The so-called "lost" Solfeggio notes refer to sets of sound frequencies used in ancient Gregorian chants, which were subsequently set aside and forgotten when newer musical forms appeared. What apparently made these chants unique was the fact that they represented electromagnetic frequencies of the Solfeggio Scale, which were thought to raise the spiritual awareness of anyone who performed or listened to this music. [Solfeggio scales are based on the key of C, instead of A. This change was put into practice by the Catholic Church, who apparently wished to embody the concepts of *subjugation* and *surrender* in their music (Bennett, 2014)]. According to recent consciousness theory, our 12-strand human DNA is presently in the process of being fully activated. This will serve to elevate the human consciousness to new levels of enlightened intelligence and evolution. During this activation process, color, sound and thought technologies can be used to reprogram human DNA with the new information. This new programming not only activates the 12 electromagnetic strands of DNA but also influences the surrounding energy layers of the DNA in such a way as to connect the human consciousness to the 12 spiritual dimensions.

The sacred solfeggio scale was "re-discovered" in 1974 by naturopathic physician, Dr. Joseph Puleo. Through a bizarre series of investigations, using the Pythagorean method of numerical reduction, Dr. Puleo deduced that these solfeggio frequencies were hidden in the biblical book, *Numbers*. He also determined that, for true healing to occur at the cellular level, and for the DNA to be re-programmed, it must first be accessed by opening up the cell membranes to be more receptive. One approach for achieving this is to activate the cell receptors (messenger RNA), which then transmits healing frequencies to the nuclear DNA. According to Dr. Puleo, the Solfeggio note frequencies open up the cells to make them receptive to DNA re-programming. The historical documentation of Dr. Puleo's work provides fascinating implications for the possibilities of establishing increasingly more advanced musical healing technologies (Hulse, 2012).

Dr. Puleo and Dr. Len Horowitz co-authored a book called, *Healing Codes for the Biological Apocalypse*. The authors provide a controversial, but fascinating, overview of the latest biological, chemical and electromagnetic weapons allegedly being used for intentional genocide and world population control. They also provide detailed information about new infections that resist conventional antibiotics such as "mad cow disease." The authors relate how scientists have discovered crystal-like structures called "prions" that appear to be spreading through fungal infected grains. They suggest that these human prion epidemics are not "natural" but may be analogous to electromagnetic frequency receivers and biological energy transmitters, deliberately designed to sicken humanity with the ultimate objective of controlling global population. The book also reveals specific "divine musical tones" which the authors claim are destined to be sung by 144,000 people – the "critical mass" necessary to establish 1,000 years of world peace (Horowitz and Puleo, 1999).

MASTER KEY 6
Earth Energies

> "Conspicuous ancient monuments
> like Stonehenge, Avebury, Carnac and the Great Pyramid,
> along with the many thousands of stone circles, earthworks and holy places
> mark the Earth's power centers all over the world.
> While ancient monuments are from another time and culture,
> the use of stone and earth to design uplifting and sacred spaces
> is just as relevant today, as it was thousands of years ago."
>
> *Chuck Pettis, 1999 - Secrets of Sacred Space*

This Key Paradigm of Future-Science Technology deals with the research, development and practical applications of ancient geomantic sciences such as Chinese Feng Shui and the global network of earth energy lines and power centers that grid Planet Earth. Fundamental aspects of earth energies include: Telluric Currents, Ley Lines, Power Points (energy vortices) and the Global Energy Grid. In the past I have been privileged to travel to many of natural earth power centers as well as visit temples and pyramids in Europe, Kuwait, Central and South America and Canada. Each of these trips was focused on experiencing the unique energies of these places first hand, viewing their art and artifacts and learning about the culture of these ancient civilizations.

With uncanny mathematical precision, enlightened architects of ancient civilizations based the design proportions and locations and of their key buildings and temples to coincide with ley lines and the intersections of ley lines (power centers). Government buildings and religious shrines were thus able to take advantage of strategic earth energy centers to extend, reinforce and sustain their power and spheres of influence in the vicinity of the cities or towns where they were located.

Magnetic Therapy for Healing and Sleep Enhancement

According to Dr. Dean Bonlie Earth's magnetic field has decreased by about 80 percent over the past four thousand years. He considers this geomagnetic field to be a critical environmental factor for the health and well-being of humans and all living things. His assumption is that, although humans have adapted over time to this loss of natural energy, there has been a corresponding loss in general vitality and the in the efficiency of the systems of the human body.

Dr. Bonlie began his career as a dentist, but over the past 20 years he developed a compelling interest in the effects of magnetic fields on the human body and the effects of magnetic supplementation to correct energy imbalances. Because of his groundbreaking research on biomagnetics he is regarded as an expert in the field of electromagnetic healing.

Taking his research to the next level, Dr. Bonlie developed and patented a powerful treatment magnet that he called his Magnetic Molecular Energizer (MME). This device is currently operating in several Advanced Medical Research Institute clinics. As of 2011 advanced clinical trials for

Phase III FDA approval for the MME were well underway. Although treatments were initially focused on neurological and orthopedic applications, patients have benefitted significantly in areas of spinal cord and brain injury, stroke impairment, multiple sclerosis, muscular dystrophy, cerebral palsy, Parkinson's and Alzheimer's disease, congestive heart failure, and bone and joint repair. Results are still being evaluated.

The MME procedure uses strong direct current electronic fields of 3-5,000 gauss. These fields are produced by two powerful electromagnets with the patient lying at the focal point between the two. The theory behind the MME device is that the human body is electromagnetic by nature, being composed of charged particles such as atoms, electrons, protons and ions. These particles interact to sustain the physical body. During treatments there is a temporary increase in the magnetic force in the atoms of the patient's body in the area of the MME focal point. This force imparts a higher velocity of some of the orbiting electrons. The increase in velocity of the electrons causes the atoms to wobble and induces a higher charge on the valence electrons that orbit around the nucleus. Both the higher velocity and increased charge serve to enhance electron transfer, the basis for chemical reactions in the body (Advanced Magnetic Research Institute, 2014).

The magnetic field of the MME device functions as a catalyst to enhance chemical reactions in the human body through the process of *electronic resonance*. Resonance is important since it improves basic body functions such as oxygen carrying capacity, nutrient assimilation, enzyme production, metabolic waste removal, free-radical reduction, tissue regeneration and an overall healing affect which spreads throughout the entire body. Resonance thus assists in repairing cellular damage, boosts enzyme production and generally enhances the immune system.

Two main factors tend to negate the effects of *natural* electromagnetic resonance: 1) The gradual decline of Earth's geomagnetic field (about 80 percent of the field strength which existed only 4,000 years ago) has reduced the energy state of the atoms in our bodies, making magnetic resonance more difficult. 2) With the rise of mobile and cordless phones, cell phone towers, smart meters and wi-fi over the past few decades, the resulting electronic smog tends to override the natural frequencies that nourish our brains, organs and tissues. This tends to bring on a state of fatigue which can lead to more serious problems like Chronic Fatigue Syndrome, which is essentially caused by a lack of natural resonance between our living tissues and the electromagnetic fields to which we are exposed to (Bonlie, 2012).

In 1991 Dr. Bonlie established a company called Magnetico, which was formed to design, patent and market a line of magnetic sleep pads. He felt his products represented a "revolutionary sleep system" which could reproduce the natural magnetic field of Earth as it was 4,000 years ago. In a recent year study, 1,375 patients were compared with age-matched controls from the CDC (Centers for Disease Control) database. Patients who used the 5-gauss Magnetico sleep pads showed a dramatic reduction in cardiac events, including heart attacks and strokes. There was also a 76 percent reduction in new cases of cancer. Ongoing research would suggest that the stronger magnetic sleep pads of 10 and 20 and 40 gauss would yield even more benefits.

Benefits for individuals who have used a Magnetico sleep pad include the following: 1) 95 percent of arthritis sufferers had less pain. 2) 91 percent of patients with sleep disorders reported sleeping better. 3) 60 percent of Fibromyalgia patients improved by 60 percent. 4) 90 percent of patients had lower resting heart rates. 5) 75 percent of patients showed increased blood-Oxygen levels. 6) Athletes using the Magnetico sleep pads reported that 80 percent of post-exercise soreness was eliminated. 7) Chronic Fatigue Syndrome sufferers reported a 75 percent improvement in symptoms. 8) PMS sufferers reported over 80 percent reduction in symptoms (Starr, 2014). The company offers a six-month trial period with a money-back guarantee. Information on Magnetico products and a video explaining how the system works can be found at their website: www.magneticosleep.com

New Future-Science Technology Applications for Earth Energies

I would like to propose that a new science called Earth Energy Technology be established. This new science would integrate earth energy principles, as they relate to the global energy grid, ley lines and power centers, with advanced medical and wellness technologies. This new science would integrate technologically advanced applications of traditional earth energy science with updated applications of "energy architecture" and "earth acupuncture." The implications for this new science are that it can be used to create subtle-energy fields, which have been shown to enhance human health, intelligence and well-being, especially in workplaces and urban environments. Other applications could include the enhancement of food production and the restoration and revitalization of our major planetary ecosystems.

MASTER KEY 7
Psychic Cleanliness and Conscious Evolution

> "Ultimate balanced health must include spiritual fitness.
> Just as Superfoods, dietary strategies and proper exercise
> are important to keep the body in shape,
> personal exploration that opens to deeper wisdom
> is necessary to keep the soul in shape.
> A sense of inner balance and deep spiritual awakening
> further accelerates mental and physical fitness."
>
> *Sam Graci, 1999 – The Power of Superfoods*

Learning to Shift Frequencies

We exist in a "sea of frequencies" that continuously interacts with our own unique personal field. Many of these frequencies and the shifts that accompany them can be extraordinarily beneficial and uplifting, as most of these beneficial frequency shifts serve to accelerate the evolution of humans and our planet. Learning to consciously shift frequencies is important for our health, emotional balance and consciousness.

Each time a frequency shift occurs our energy fields are correspondingly modified. We simply need to adjust our perspectives and operational protocols accordingly. Often, adjusting to each shift tends to cause our physical, emotional and mental bodies to react in response to being moved from their natural comfort zones. Electronic smog can create unwanted frequency shifts. This is why an awareness of these human-created frequencies and how they impact our biological energy field is so important.

Cleaning Our Energy Fields and Shifting into an Interdimensional Quantum Consciousness Framework

As we continue to expand our thinking into the quantum field and work toward developing our own methods for using and experiencing it, we can continue to evolve this concept further by expanding our thinking perception into the higher dimensions. This shift requires the psychic clearing of embedded past programming accumulated from our family, educational and social interactions. This involves *both* the physical and higher vibrational bodies and requires time and perseverance. The process can be facilitated by outside professionals, who can help us identify these embedded bits of programming. These embedded programs can be compared to boulders in a stream that interfere with the water's natural flow. These aberrations have been referred to in the literature as "engrams" or "holodynes." These bits of "corrupted programs" also extend back through our family lines and past-life experiences. Dr. Vernon Woolf has developed this concept through his books and courses in Holodynamics, where he teaches others how to identify and manage these holodynes so they cease being blockages, and instead can be evolved to their full potential to support the expression of an individual's highest self (Woolf, 2004).

In addition to this ongoing process of cleaning and clearing of our energy bodies, we need to learn to transcend the ego-self and expand our consciousness into the higher dimensional realms. This requires disciplining our awareness to work above the ego-self, a major point of attachment for controlling programs. Working above the ego-self begins with the recognition of *when* the emotions and past programming attempt to influence our thoughts, words and actions. Although this requires time and discipline the rewards are exponential in terms of the effort invested. The ultimate objective is to expand our consciousness into the higher realms where we can contact, interface with and evolve into our highest potential self. From a higher vibrational perspective we can see problems in a different way and often turn them around to become opportunities. Since our higher self exists in a no-time/no-space state it can perceive any problem or situation from the highest perspective and shift potential problems into resolvable solutions.

Psychic Cleanliness: A Consciousness Evolution

Hygiene is the science which deals with the perseveration of health of the physical and higher vibrational bodies. In the larger context, it relates to the general cleanliness of our body, homes, workplaces and public areas. From a paraphysical perspective hygiene begins with cleanliness of the physical body and extends to include the higher vibrational bodies.

During our daily activities the electromagnetic field that surrounds and permeates our physical body tends to develop ridges, barriers and tears that let in inflows of outside energy. These inflows from outside sources may be of human origin in the form of verbal, mental or psychic thought forms. They can also originate from manmade electromagnetic fields or from planetary, solar or extraterrestrial radiation sources. When such outside forces interact with an individual's electromagnetic field, blockages tend to form within the body's normal energy patterns. When these blockages persist for prolonged periods of time they tend to form more permanent structures that are called "aberrations." Unless these aberrations are periodically removed, they can remain embedded in the energetic body's electromagnetic field, distorting the natural energy patterns and causing problems in the physical body. In the same way negative thought forms generated by anger, hate, anxiety, impatience, etc., when held over prolonged periods of time, can also become deeply embedded in our electromagnetic fields. Eventually, these negative elements begin to manifest at the physical level, resulting in disease conditions that attack the physical body at its weakest points. In a similar manner, injuries to the physical body may also cause aberrations in the personal electromagnetic field and thus affect the body's energy flow patterns to varying degrees.

Special techniques are available for cleaning the physical and higher vibrational bodies, and for removing aberrations from the human electromagnetic fields. These practices have existed for centuries in the Orient and include acupuncture, acupressure, yoga, smudging and sweat lodge ceremonies. The objectives of these ritual therapies is to remove, neutralize and counteract the effects of aberrations and negative inflows, while simultaneously clearing, re-aligning and restoring the natural energy patterns of the human energy system. Through specific applications of Future-Science Technology, electronic devices and psychic techniques can also be used to cleanse the physical and extra-physical bodies and to sustain them in a state of healthy energy flux. Once an individual's energy fields are aligned and stabilized, previously existing problems will tend to automatically correct themselves and the body will sustain a state of high energy which keeps the immune system healthy and prevents the occurrence of disease. The practices a meditation, yoga and the soft martial arts, combined with electronic consciousness devices (mentioned in other sections of this book) can also increase physical and mental energies and enhance fitness and athletic conditioning.

Overcoming Tribal Beliefs Embedded in the Subconscious Mind

Tribal beliefs are unconscious programs that shape our family, religious or educational belief systems. Examples would include such statements, as "I am not worthy," "Boys are smarter than girls," or "We have to work as hard as possible and earn lots of money." Such beliefs are energy-draining and not in alignment with our higher selves. For most of us these tribal beliefs are so deeply embedded we consider them to be basic truths. As such they tend to form the foundation for most of the emotional and social aspects of our lives.

Tribal beliefs reside in the first chakra where they tend to erode our sense of security and our basic desire to live up to our full potential. These deeply embedded beliefs affect our health and well-being because they create a conflict to either conform to tribal laws or listen to our higher selves. When we choose to reject the restrictions of tribal law and follow our higher calling, it is common for basic survival issues to surface. Often this kind of conflict causes blockages in the root chakra, where the energies are linked to issues of survival. Survival issues most often tend to override issues of sexuality, which also reside in this first chakra. These survival issues include our sense of security in the world, our desire to be alive in the present and the quality of relationships with our family, spouses and others. Depending on the basic nature of our tribal beliefs they can be either beneficial or negative. Whatever the case, tribal beliefs reside in the unconscious until we make a deliberate effort to uncover and acknowledge them. Once we do this it is an easy matter to change them.

The power exercised by toxic beliefs in shaping and controlling the masses is nothing short of shocking. Fortunately, the flip side of this is that we also can gain many benefits in terms of healing, positive energy and enlightenment by rising above these culturally imposed restrictions and shifting into receiving new guidelines from a higher source. Other "shadows" can also be caused by traumatic events in our lives. Uncovering and acknowledging these shadows is critical to our physical, emotional and spiritual health, because these shadows can prevent us from healing and evolving into our full potential selves. Unless we are willing to be fully committed to changing our outdated belief systems and healing, we will never get well and stay well.

MASTER KEY 8
Tesla Technology

> "My method is different.
> I do not rush into actual work.
> When I get a new idea, I start at once building it up in my imagination, and make improvements and operate the device in my mind.
> When I have gone so far as to embody everything in my invention, every possible improvement I can think of, and when I see no fault anywhere, I put into concrete form the final product of my brain."
>
> *Nicola Tesla*

As a kid of five or six I would often visit my paternal grandfather. He was chief engineer at the Central Maine Power Plant in Rumford, Maine. He would take me to this amazing place where he lorded over a team of electrical engineers. I vividly remember standing on the balcony overlooking the penstocks, which directed rushing water down into the massive turbines that drove the huge red electrical generators, positioned on top of them. Sometimes he would take me down to stand on the steel grates on top of the generators. I well remember the overwhelming sense of awe in feeling these giant machines vibrating under my feet. My grandfather lived to the age of 97. I often regret not asking him about Nicola Tesla's work, as I am certain he would have had some interesting comments to share with me.

Tesla Technology:
A Quantum-Field Tool for Healing

Yugoslav-born scientist Nicola Tesla (1856-1943) immigrated to America, where he began working for Thomas Edison in New York. During his erratic career he designed and developed alternating current, electric motors and a system for generating and distributing alternating electric current that established the basis for our modern electrical systems.

Tesla's discoveries were so basic, numerous and so far ahead of his time, that he has often been depicted in the literature as "the prodigal genius." His breakthrough concepts in electrical engineering were responsible for the first tuned radio circuits, discoveries in gaseous and fluorescent lighting, the use of high-frequency electronics in medicine and the first demonstrations of robotics. The series of air conical transformers, which he designed and built, have become known as "Tesla coils." Since this type of coil depends on the resonant electrical vibrations produced by the interaction of the coil with a capacitor, the production of very high voltages became possible.

Interestingly, Tesla's thought process apparently operated at a very high level (within the quantum field), so he was regarded by the historians and intellectuals of his time as a super-genius. He perceived the natural world in terms of vibrational phenomena and understood the relationships between the resonant circuits he developed and the acoustical resonant principles of physical matter. His thinking and scientific inventions were thus focused on the principles of *mechanical vibration* and *resonance*. Tesla apparently developed a form of "etheric vision" which allowed him to image his experimental designs in his mind. Within this meditative state, he was purported to be able to assemble the components and de-bug his inventions before bringing them down into the physical reality.

Many of Tesla's devices – especially high-frequency, high-voltage transformers and harmonically tuned resonating coils – were designed to operate *without* direct electrical connections and were sourced from quantum-field/zero-point technology.

The Multiwave Oscillator: A Healing System Based on Tesla Technology

One device based on Tesla's technology is the so-called Lakhovsky Multiple Wave Oscillator, or MWO. Basic research for the MWO was accomplished prior to World War II. Lakhovsky's theory was based on the fact that most cells in the body contain nuclear filaments of highly conductive material, which are surrounded by insulating media (the DNA-RNA Complex). The helical configuration of this DNA/RNA complex is analogous to a coil, which has the ability to react as a tuned circuit when its resonant frequency is stimulated by an external oscillating electrical coil. Lakhovsky further postulated that by thus exciting the cell nuclei, a "charge" could be induced through the principle of *electromagnetic induction*. Accordingly, the energy level, or "vitality," of every cell within the field could be raised simultaneously. Since each cell of a biological entity is of slightly different physical dimensions, the exciting wavelengths were designed to be multiple, and spanned a broad frequency spectrum.

Multi Wave Oscillators were once commonly used for medical healing throughout France, Germany and Italy. These devices were also tested in New York hospitals, where they were said to demonstrate cure rates of up to 98 percent in cases of terminal cancer, arthritis and conditions considered "hopeless" by conventional medicine. However, due to the unorthodox design and quantum field effects, combined with resistance from the American medical establishments, MWO technologies were abandoned and all but forgotten in the United States. What *is* significant about these medical devices is that they utilized electronically generated frequencies to resonate with living cells within the quantum field, and thus facilitated a healing process, which proved to be successful in healing many otherwise incurable conditions.

In the context of Future-Science Technology, the basic designs of the Lakhovsky Multiwave Oscillator and other Tesla devices have been refined and updated with contemporary solid-state electronics. In addition to their healing applications these devices can be used to generate biological enhancement fields, which stimulate the physical body's natural potential for positive self-expression. Such applications of Tesla Technology are fundamental aspects for creating and sustaining vibrational fields, which trigger the maximum expression of an individual's highest potential self.

The HAARP Project:
An Example of Tesla Technology in Action

HAARP stands for High Frequency Active Auroral Research Program. The program is funded jointly by the U.S. Air Force, (ONR) the Office of Naval Research, DARPA (Defense Advanced Research Projects Agency) and the University of Alaska. The project operates mainly from the HAARP Research Station located on an Air Force Base in Gakona, Alaska. By 2008 this program had spent about $250 million taxpayer dollars for construction and operational costs. The basic components at the Gakona facility include: An array of 180 phased antennas, various scientific analytical instruments for signal monitoring purposes and three atmospheric radar systems (Wikipedia, 2014). Other facilities are located in Australia, Norway, China and Russia. The following website provides a list of HAARP-related installations: (Before Its News, 2011).

The conspiracy issues surrounding HAARP should not be taken lightly, as among its critics are respected scientists who believe that the electronic warfare capabilities of these installations are designed to achieve the U.S.

Military's goal of "full spectrum dominance" by 2020. Major aspects of the project have been kept secret "for reasons of national security" (Burks, 2010). Perhaps in order to avoid negative publicity, the HAARP Facility was closed down in May of 2013, allegedly due to lack of funding and failure to meet the local EPA air quality requirements (Paul, 2013).

Although HAARP Technology is most often associated with Tesla Technology, much of the research involves ELF (Extremely Low Frequency) and VLF (Very Low Frequency) waves, which are critical for naval operations with regard to underwater communications, navigation and torpedo guidance.

It is not only conspiracy theorists that are deeply concerned about HAARP. The European Union referred to the project as a "global concern" and subsequently passed a 1999 resolution which called for additional information as to the project's legal, ethical and environmental consequences (Burks, op. cit.). Other HAARP critics include Dr. Nick Begich, author of the groundbreaking book, *Angels Don't Play this HAARP* (Begich and Manning, 1995) and his DVD, *HAARP – The Update* (Begich, 2006).

For basic background information, CBC and The History Channel produced two excellent documentary films. They can be found at: (CBC News, 2008; History Channel, 2013; YouTube, 2010). Although a comprehensive treatment of HAARP technology is beyond the scope of this book, it appears we are dealing with very powerful technology, which has far-reaching implications for humans and our global atmospheric commons. Here are two quotes from the History Channel documentary, which highlight the potential negative aspects of HAARP: "Electromagnetic weapons…pack an invisible wallop hundreds of times more powerful than the electrical current in a lightning bolt. One can blast enemy missiles out of the sky, another could be used to blind soldiers on the battlefield, still another to control an unruly crowd by burning the surface of their skin. If detonated over a large city an electromagnetic weapon could destroy all electronics in seconds. They all use directed energy to create an electromagnetic pulse." The documentary also states that, "Directed energy is such a powerful technology it could be used to heat the Ionosphere to turn weather into a weapon of war. Imagine using a flood to destroy a city or a tornado to decimate an approaching army in the desert. The military has spent a huge amount of time on weather modification as a concept for battle environments. If an electromagnetic pulse went off over a city, basically all of the electronic things in your home would wink out and they would be permanently destroyed."

For those who still doubt the potential of HAARP technology, a quote from New Zealand's leading newspaper, *The New Zealand Herald,* published the following statement: "Top secret wartime experiments were conducted off the coast of Auckland to perfect a tidal wave bomb, declassified files reveal. United States defense chiefs said that if the project had been completed before the end of the war, it could have played a role as effective as the atom bomb. Details of the tsunami bomb, known as *Project Seal*, are contained in a 53-year-old document released by the Minister of Foreign Affairs and Trade." Researchers have also questioned the possible use of HAARP Technology to generate and steer hurricanes like Katrina, to trigger earthquakes and tsunamis in Haiti and Japan and to create floods and droughts or bring rain to parched regions.
(www.youtu.be/3uqRP4aQjDA?list=PL376E627EAD97CF4B and www.examiner.com/article/haarp-caused-japanese-earthquake-nuclear-bomb-video).

There are other researchers who claim there is a major link between HAARP and the chemtrail phenomenon, based on the fact that both technologies involve manipulating the energies in Earth's atmosphere (Otterwalks, 2013). There is also emerging evidence to support claims, which attribute interferences in bird migrations and bee mortalities at least partly to HAARP technology (Cramer, 2007).

The following links offer videos and satellite documentation, which suggests that the torrential flooding, which occurred in Alberta, Canada in 2013, was linked to geoengineering activities (Gotz, 2013; Webre, 2013).

In closing this Pandora's Box I feel compelled to add the following footnote: I was confidentially informed by an individual who claimed to have worked for the HAARP Project, that the 6.7 magnitude Northridge Earthquake of January 17, 1994, which killed 57 people and destroyed buildings and freeway overpasses at a cost estimated at 20 billion dollars, was triggered by a U.S. Navy submarine located in one of the undersea canyons of the California coast. I have no reason to doubt this individual's credibility as well as information recently obtained from other unnamed individuals that weather engineering has been employed by the U.S. and its allies to disrupt the Winter Olympics by creating higher than normal temperatures (Taylor, 2014).

MASTER KEY 9:
ELF Electronics

> "Our lump of rock is anything but inert.
> It sizzles and crackles with energy; pulsing, itching and breathing like a living thing; responding directly to changes in itself and in the environment.
> And we ride these waves like veteran surfers, dealing instinctively with their fluctuations, anticipating changes in frequencies which lie beyond the limited scope of our usual sense organs. We learn to read between the lines.
> We resonate in natural sympathy with our planet."
>
> *Lyall Watson 1988 – Beyond Supernature: A New Natural History of the Supernatural*

Schumann Resonance: The Pulse of Planet Earth

In 1899 Nikola Tesla discovered that Earth had an innate resonant frequency of approximately 8 Hz (cycles per second). Although Tesla documented his discovery, it was not scientifically validated until 1953, when Professor W.O. Schumann of the University of Munich discovered that Earth's ionospheric cavity produces specific pulsations. Subsequent research confirmed that Earth resonates at 7.83 Hertz. This frequency was named, The Schumann Resonance. Schumann resonances are sets of spectrum peaks in the extremely low frequency (ELF) range of Earth's electromagnetic field spectrum. These peaks are electromagnetic pulses which form standing waves equal to the circumference of our planet. These ELF waves are continuously generated by lightning discharges within the cavity between Earth's surface and the ionosphere, which functions as a closed wave guide.

Following the publication of Schumann's research, a physician named Dr. Ankermueller made the connection between Schumann Resonance and brain wave frequencies. He contacted Dr. Schumann, who asked one of his graduate students to investigate the phenomenon. Subsequent studies confirmed the relationship between the Earth's resonance and human brainwaves. Dr. Wolfgang Ludwig, researcher in Earth-mind relations, then set out to define the specific frequencies that exist in a healthy environment. Ludwig found that Schumann waves could easily be measured within natural environmental settings, but were impossible to measure in urban environments because man-made electromagnetic signals interfered with and blocked out the natural Earth frequencies.

In 1963, Professor Rutger Wever of the Max Planck University built an underground bunker that screened out the Earth's natural Schumann frequencies. In a series of experiments, student volunteers lived in the bunker for up to four-week periods. Research data showed that when Schumann Resonances were filtered out of the bunker the students' physical and mental health suffered. They experienced stress, emotional imbalances and migraine headaches. After brief exposures to the 7.83 Hz frequency from an electronic frequency generator later installed in the bunker, their health and mental attitudes quickly stabilized. This research suggested that human beings were not well adapted to live in the vicinity of strong electromagnetic fields, but require Earth's natural resonance to achieve good health and well-being (www.musicforyourmind.com, 2012).

Due to the rapid advance of technology, it seems we have created our own "urban bunkers." We have succeeded in cutting ourselves off from the Earth's natural pulse by our own inventions and conveniences. Is it any wonder that stress, depression and anxiety disorders are more abundant today than ever before in history? For more information on this subject I would suggest the eye-opening video, *Resonance – Beings of Frequency*, a comprehensive documentary about the dangers we face in living in our sea of manmade frequencies (Poppe, 2014).

Entomologist and Agricultural scientist, Dr. Philip Callahan, approaches Schuman Resonances from a slightly different perspective. Callahan's extensive research on earth energies, paramagnetic soils and biological responses to electromagnetic frequencies suggest that the human brain, as well as all life forms on Earth, interact with these natural earth energies (Callahan, 1995). I have been privileged to attend some of Phil's lectures and spend quality time with him. I personally regard him as one of our unsung scientific heroes who should be nominated for a Nobel Prize for his amazing scientific achievements.

Any mass, such as a mountain, building or monument alters the vibrational characteristics of that region – just as placing one's finger on a drumhead alters the basic vibrational pattern. Through the process of resonance this pulse is reinforced and amplified. By combining ancient scientific principles with modern technology these natural earth energies can be channeled, focused, stored or re-directed to increase agricultural and livestock production, stimulate healing and rejuvenation and create fields of peace to enhance virtually all aspects of biological systems.

ELF Waves and Human Brain Waves: How They Interact

Human brainwaves are divided into four basic frequency ranges called Beta, Alpha, Theta and Delta. Beta waves typify our normal waking state and range from 14 to 30 cycles per second (cps). Alpha waves predominate during daydreaming or meditation states and range from 8-13 cps. Theta waves dominate during deep states of meditation and range from 4-7 cps, while Delta waves are characteristic of deep sleep or the deepest levels of meditation where brainwaves fall below 4 cps. The Delta state is thought to be the level of the Universal Consciousness, since it allows us to operate outside of normal space and time (i.e. within the quantum field). Thus, normal waking consciousness is typified by brainwaves operating in the Beta Range; while Alpha, Theta and Delta are referred to as "altered states of consciousness." Small children and animals function mainly in the Theta, Alpha and Delta states of mental activity, while adults operate mostly in the Beta frequency range (Brain Waves, 2010). The ELF Phenomenon has two main aspects that are paramount to the health of our planet and of all living things: 1. The Natural Earth Pulse and its interrelationship with the biological fields of all life-forms; and 2) Electronically Propagated ELF technologies such as HAARP and similar technologies. As mentioned previously, global military forces use ELF waves for underwater communications. From the time when primitive life-forms first began to evolve, planetary ELF waves have been a basic formative force, necessary for the growth, maintenance and evolution of life. From this relationship between the natural ELF fields and the electromagnetic bio-fields associated with living organisms, a delicate bio-planetary balance has evolved.

Through the use of specialized electronic ELF-field generators, the energy patterns associated with the brain and other biological systems can be entrained, suppressed or positively stimulated. Such techniques can be used to benefit living organisms in many different ways. Examples of areas where such bio-enhancement techniques might be employed include healing, regeneration, education, sleep therapy, mental alertness, emotional stability, attitude, athletic training and fitness. Similar benefits can also be realized for non-human species in agriculture, forestry, aquaculture and animal husbandry.

MASTER KEY 10:
Radionics: A Quantum Technology for Human and Planetary Healing

> "Any sufficiently advanced technology is indistinguishable from magic."
>
> *Sir Arthur C. Clarke*

Radionics: A Quantum-Field Healing Technology

The Science of Radionics deals with the interactions of matter, energy and consciousness. Radionics is based on the discoveries of Dr. Albert Abrams. In 1910, Dr. Abrams and subsequent investigators developed electronic medical devices, which combined a radio-frequency amplifier, a variable condenser tuning system and a non-inductive resistance plate. These devices were apparently able to create a "consciousness linkage" between the instrument and its operator. A corresponding patient-instrument link was created by placing a photograph or organic sample from the patient (blood, urine, hair, saliva) in the tuning well of the instrument while in the receiving mode. A plastic rubbing plate with pick-up coil provided the means for aligning the operator's energy field with the instrument and patient. When the emanations from the tuning component on the instrument matched the numerical frequency of the sample from the patient, resistance was encountered on the rubbing plate (essentially a form of dowsing). A set of numerically calibrated dials on the instrument provided a number set for each situation. The instrument was then switched over to "broadcast mode," thus broadcasting these same frequencies to cancel out the disease condition initially detected by the device.

Through the principle of *resonant interaction* it was possible to catalog a series of vibrational combinations, represented by exact numerical settings, on the radionics instrument for each particular disease condition or patient. Although these instruments made no sense when compared to conventional electrical circuitry, there are numerous well-documented cases in the literature of patients considered incurable by conventional medical practices being successfully cured through radionics treatment. Perhaps the most amazing aspects of this kind of healing technology are that it functions non-locally in a quantum context such that the patient and healing device can be physically separated by thousands of miles.

Despite the hundreds of documented cures that were accomplished over the years through radionics technology, the unorthodox nature of the technology generated tremendous resistance from the mainstream medical establishment. The most controversial aspect of radionics lay in the fact that the instruments appear to function within a quantum modality which is independent of conventional time-space coordinates. In addition, the fact that radionic diagnosis and healings were routinely accomplished using a photograph of the patient, placed in the tuning well of the instrument, simply exceeded the limits of traditional medical practice.

The scientific explanation for radionics is based on the concept that a photograph or sample from the patient represents a holographic version of the patient and their condition at a given time-point. In the case of the photograph this holographic representation is embedded within the crystalline matrix of the photochemical emulsions. Thus, when a photo is placed in the electronic field of the radionics instrument a specific interaction occurs which relates directly to the vibrational patterns of the patient.

Some of the best-documented applications of radionics have been in the field of agriculture, since it posed no threat to the medical establishment. Agricultural radionics thus quietly continued to be developed and refined. Radionics has proved effective for eliminating insect pests without insecticides and balancing out soils to increase crop yields without the addition of chemical fertilizers. This was an important breakthrough, since when approached by radionics practitioners; farmers were unwilling to pay unless the treatment yielded undeniably positive results.

Radionic treatments have been validated in the field of veterinary medicine, where dogs, cats, racehorses and dairy cattle have all been treated. Treatment of animals seems to have been even more successful than with humans. Since animal patients have no preconceptions, they accept these treatments unconditionally – unlike human patients who sometimes exert mental resistance that interferes with the treatment results.

Radionic Aquaculture: A Futuristic Paradigm for Intensive Food Production

Aquaculture is the science of raising aquatic organisms such as algae, crustaceans or fish under controlled environmental conditions. Indigenous cultures have practiced primitive aquaculture for millennia, but only since the late 1960's has aquaculture become a significant factor in global food production.

To my knowledge, the use of radionics technology has not yet been openly used in aquaculture, however there is no reason why the successes achieved in agriculture and veterinary medicine should not be transferable to all aspects of aquaculture. I thus propose to integrate radionics technology into aquaculture sciences. My plan is based on my own past pioneering ventures into fresh and salt water shrimp aquaculture and a unique perspective which perceives an aquaculture pond or intensive culture system as a single integrated biological entity, with the culture species representing interactive organic products of the system. For example, radionic treatments could be substituted for commonly used antibiotics to remove disease organisms, sustain pH values and generally increase the efficiency of these systems.

Based on agricultural radionics models it should be possible to direct specific frequencies into the systems to improve growth rates, survival, taste and nutritional value.

The integration of radionics into aquaculture science would represent a significant step forward for improving our strategic food resources. The implications of this scientific advancement should be apparent for all types of aquaculture species, especially with regard to increasing food production in arid lands, third-world countries and for growing fresh food for urban areas of the world.

Some Proposed Futuristic Application for Radionic Technologies

In addition to its medical, agricultural and veterinary applications I am convinced that radionics technology, in its present format, represents only the tip of the iceberg, since this technology could also be applied to many aspects of science, professional and personal life. Possible applications could include the fields of education, business, sports, global ecology, weather modification and space sciences. This offers exciting promises and possibilities – especially if existing radionics devices are updated with the latest electronics advancements. Although special training and a certain degree of psychic sensitivity are required, the basic equipment is relatively inexpensive when compared with conventional scientific or medical equipment.

We no longer need to fear radionics because of its unique quantum nature. Instead, we should embrace it, continue to validate results and then describe the new physics behind this science – similar to the way Dr. Tom Bearden wrote his groundbreaking book, *Energy from the Vacuum: Concepts and Principles*, to establish and validate the basic principles of zero-point energy (Bearden, 2002).

∞

A Modest Proposal: A Future-Science Approach to Planetary Management

The major problems we face as a species relate directly to imbalances in the global ecosystems. These problems include: overpopulation, famine, disease, deforestation, desertification, air and water pollution and weather-related disasters. If it were possible to offset these negative trends on national and global scales, the impact of human technology on the global biosphere would begin a series of positive shifts into a new, sustainable framework for humans and the global biosphere.

The author thus proposes a revolutionary new program of research and development, which focuses on developing advanced approaches and technologies into a practical framework for planetary-scale ecological programming. The basic radionics technology required for such a project has been available since the 1940's, but its applications to date have been limited to medical, veterinary and agricultural sciences. By updating the instrumentation to include planetary feedback and by modifying the basic applications, it should become possible to halt and reverse presently destructive ecological trends on a global scale and within relatively short time frames.

The implementation of this proposed project does not require vast manpower or funding resources when compared with conventional government or military projects. The program would utilize updated radionics technology, coupled with existing satellite and communications technologies. It would be designed to function within the operational matrix of Future-Science Technology for the mutual benefit of all life on Earth.

MASTER KEY 11
New Futuristic Technologies

> "Research and development on advanced, unconventional renewables
> has thus far been suppressed by very powerful interests.
> However, new policies could quickly reveal that vacuum (zero-point) energy,
> cold fusion, advanced hydrogen and water chemistries, and other novel approaches
> can lead to a quantum leap in having a clean, cheap, safe, decentralized energy future.
> New energy technologies may be our only lasting hope to reverse the global climate
> crisis. We must therefore support the responsible development of these sources
> for a sustainable future Earth."
>
> *Brian O'Leary, 2008 – The Energy Solution Revolution*

Non-Conventional Technologies as Change Agents for Human and the Global Biosphere

Humans everywhere now find themselves facing daunting environmental and social problems on local, national and global scales. Although conventional science and technology has come up with some solutions for these problems, in most cases practical solutions have failed to materialize. The good news is that there *are* non-conventional technologies existing today that can reverse destructive environmental trends and repair existing damage.

Throughout the 20th Century independent researchers have developed numerous alternative technologies. These "orphan technologies" have proven effective in areas of communications, electrical power generation, transportation, agriculture and medicine but have either been ignored or hidden from public view. The common denominator for these orphan technologies is that they function *outside* the boundaries of conventional scientific thought. Thus, it has been impossible to validate them within the established scientific framework without changing the framework itself. Despite that fact that definitive, replicable results have been demonstrated, scientific and corporate alliances with major vested interests in maintaining the *status quo* have worked relentlessly to discredit these pioneering scientists – often isolating them from their more conventional scientific peers.

Historically, the mainstream scientific community has invested a major portion of its energies toward protectionism and self-perpetuation. However, every so often the bastions of mainstream science have become so overwhelmed by shattering glimpses of the obvious that they have been forced to update their existing paradigms. Galileo and Charles Darwin provide historical examples of this phenomenon. Modern examples of ongoing scientific paradigm clashes include the unreasonable reactions of academic and scientific institutions against early researchers in the field of cold fusion [Their work was subsequently validated by the U.S. Office of Naval Research]. The same is true in the field of zero-point energy where established physical laws break down, and the barriers between classical science and consciousness science become blurred.

In response to this unprecedented crisis the author offers viable solutions through Future-Science Technology – solutions that combine the best of conventional and alternative scientific approaches. This new perspective integrates conventional scientific techniques with advanced psychic technologies and contains practical, environmentally sensible protocols, which stretch the limits of conventional scientific thought.

To establish enlightened pathways into a new third millennium mindset, a concerted effort thus needs to be initiated by world leaders to support alternative technologies and to fund their development and integration into society. Funding support from government, academic, scientific and corporate sectors are another key factor for bringing these new technologies into practice. Also important is the fact that new technologies create new jobs and provide new investment opportunities. If this future-science perspective can be actively embraced and woven into the existing global social fabric, resistance from established entities *can* be turned around so that everyone concerned becomes part of the program.

It is time for the human species to pick its knuckles off the ground and abandon its present self-centered values of *ego-gratification*, *me-first*, and *over-consumption* – replacing them with positive values such as *high integrity*, *self-respect*, *self-responsibility* and *a concern for the welfare of others*.

Within the context of synergistic support and international cooperation, egocentric barriers can be dissolved and the demons of fear, greed, competition and insecurity of the past can be left behind. The thrust of new-millennium science should thus be to demonstrate that it *is* possible for alternative scientists to establish synergistic relationships with conventional scientific institutions and to develop eco-sensible new technologies, which combine the best of both worlds. Pathways of natural transition can then be charted, so established interests can discover profitable new ways to save face and prosper more efficiently. It is equally important that alternative theories be grounded into practical working models and applications, which demonstrate irrefutable results – leaving theoretical models and hard scientific proof to be debated and defined at a later time. Major funding resources can function as the fundamental driving force for facilitating the integration of alternative science into society.

Within the spirit of third millennium future-science, the self-serving attitudes of ivory tower scientists can thus be shifted to embrace practical, outcome-based applications, which focus on improving the general quality of life. In this same context global military funds can be redirected into forming an environmental Earth corps dedicated to enforcing environmental regulations, providing disaster relief, protecting natural resources from environmental plundering, implementing clean-up and restoration projects and generally providing technical and manpower support where and when it is needed.

A Car that Needs Refueling Every 100 Years

A new engine that is currently being developed in the U.S. could result in a car that only needs fueling once every century. A company called Laser

Powered Systems is building a unique turbine engine. An electrical generator will be connected to a special thorium laser. Thorium is a mildly radioactive element that is one of the densest materials on Earth. Since thorium is so dense, it is also highly efficient in producing energy. To put this into perspective, a single gram of thorium yields more energy than 7,400 gallons of gasoline. Thus, eight grams could theoretically power a car for 100 years.

Laser Powered Systems already has plans to bring its engine into mass-production. The company envisions a future where their thorium turbine engines could power a significant percentage of the cars on highways worldwide, effectively reducing air pollution from hydrocarbon fuel emissions worldwide. The most recent version of the company's new engine weighs about the same as an older conventional V8 engine, so it will fit under the hood of many existing vehicles.

This is not the first time thorium has been considered as an ideal fuel for automobiles. For example, in 2009 General Motors unveiled their "Cadillac World Fuel Concept." GM projected a 100-year refill period for their vehicle and planned to include a free maintenance plan for all systems, but the concept failed to produce a working prototype.

Laser Powered Systems has an optimistic outlook. Their website promotes their promise of emissions-free power. This, they claim, would represent a new sustainable green technology, which could also help revitalize the American automotive industry (Pincott, 2014).

The following video provides an overview of the new thorium vehicle concept: www.youtube.com/watch?v=68A_HPYGdlk
The company website can be found at: www.laserpowersystems.com

3D Printing: An Amazing New Technology Already Up and Running

One new technology, which has recently emerged into the public domain, is 3D printing, a technology that creates three-dimensional solid objects from a digital model. The process is accomplished using additive processes, where an object is created by laying down successive layers of material. Although the process cannot yet replace traditional machining techniques, it represents a quantum leap when compared with traditional subtractive processes, which rely on removal of material by methods such as cutting, milling and drilling.

3D printing is performed by a materials printer, which uses digital technology. The process involves the application of virtual designs from computer-aided design (CAD) or animation modeling software. It transforms the design into thin, virtual, horizontal cross-sections then lays down successive layers until the model is complete. The machine reads the design software and then "prints" successive layers of liquid, powder or sheet material to build up the model from a series of cross sections. These layers are then joined together or fused automatically to create the final shape. The major advantage of additive fabrication is its ability to create almost any shape or geometric feature. Although the technology was developed in the 1980's, it has been refined to produce toys, jewelry, footwear, architectural models, automotive and aerospace components, dental crowns and medical prostheses (Wikipedia, 2012).

Other fascinating examples of new ways that 3D printing is touching our lives include the creation of a custom 3D printed robotic exoskeleton for 4-year-old Emma Lavelle by researchers at Philadelphia's Alfred I. DuPont Hospital for Children. Emma's so-called "magic arms" are helping her overcome a congenital defect, which limited her joint mobility and weakened the associated muscles. When Emma outgrew her first exoskeleton, researchers simply created an enlarged version of the original. Since the success with Emma, over a dozen disabled children have been fitted with 3D printed exoskeletons (Coldewey, 2012).

3D printing technology is no longer just an idea. The technology has literally exploded over the past few years and is rapidly becoming increasingly more visible in the public domain. Basic 3D printer kits are now available for as little as $250. Although such home or shop kits are useful for making a variety of household objects, the most common use is apparently for fixing broken things. Although learning how to use CAD (Computer Assisted Design) programs is admittedly a daunting process, with more 3D computer plans becoming available, downloading 3D printer programs is easy and usually free. 3D printing thus offers infinite possibilities for changing our lives. This amazing new technology has moved from the realm of the imagined into the real world and continues to develop exponentially (Huffington Post, 2013).

In 2012 a company called MakerBot Industries opened its first 3D printing store in Manhattan. The first of its kind, the store showcased its second-generation replicator 3D printer. The unit sells for about $1,900 and is already in use by artists and hobbyists who use it to create small objects such as prototype models, customized toy soldiers and doll house furniture (Diep, 2012). Scientists have also created a 3D metal printer, which allows anyone with access to a welder and the internet to make their own metal tools or replacement parts. The MakerBot unit sells for about $1,200. at the time of this writing (Linux Today, 2013).

Unique new applications of 3D printing technology continue to surface rapidly. For example, a company called 3D Concepts has successfully produced what it claims to be, "the world's first 3D metal printed gun," which has already fired over 50 rounds. Construction of the semi-automatic handgun employed a process called laser sintering to fashion the 30 individual parts for the completed gun. Regarding public safety, 3D Concept's Director of Marketing, Scott McGowan states: "There are barriers to entry that will keep the public away from this technology for years. These include a prohibitively high cost for the equipment involved and the expertise involved to actually pull off the printing" (Welch, 2013).

A New 3D Printer that Uses Embryonic Stem Cells

Researchers at the University of Edinburgh have developed a new 3D printer that has the potential to create human tissues for testing new drugs and possibly even growing new organs. Human embryonic stem cells have traditionally been obtained from placentas that are discarded at birth. These cells are unique in that they contain the potential to develop into virtually any type of cell in an adult human or animal, including brain, muscle or bone cells. This quality makes stem cells ideal for applications in the field of regenerative medicine, which involves repairing or replacing, damaged cells, tissues and organs. To accomplish this, the stem cells are placed in a nutrient solution containing specific biological components, which guide the undifferentiated cells to develop into the desired tissue (Gismodo, 2013).

Building a House in 20 Hours with 3D Printing

University of Southern California Professor, Behrock Khoshnevis, is an out-of-the-box thinker with a background in engineering, robotics and computer-aided design (CAD). Professor Khoshnevis was troubled that our 21st century world is still rife with poverty-stricken slums and corrugated tin shacks. He thus set out to improve the concept of house construction by making it more accessible to everyone. Conventional house construction is presently a slow, labor-intensive and dangerous process that usually runs over-budget. Unlike the automotive or airline industries that embraced automated production methods to accomplish routine construction tasks, housing still requires extensive manual labor.

In searching for a new approach to house construction, Professor Khoshnevis came up with the idea of using 3D printing. He applied his engineering experience to scale up 3D printing and adapted it for housing construction. He calls his new process "Contour Crafting." Khoshnevis envisions building entire neighborhoods using contour crafting technology and claims it can be done at a fraction of the cost, and within a much shorter timeframe. He projects that 3D printed houses will cost 25 percent less than traditional houses, and that labor costs can be reduced by half. In terms of construction time, he states, "We anticipate that an average 2,500 square foot house can be built in about 20 hours from a custom design."

The contour crafting process can be summarized as follows: A CAD design is sent to a large 3D printer that has been set up over the building site. The printer lays out the concrete-like foundation of the house through nozzles that can move anywhere on the building site. The house is constructed one layer at a time and reinforced with various materials (rebar, electrical, plumbing and communications lines). The supersized 3D printer applies a concrete-fiber polymer mix, which is over three times as strong as ordinary concrete. One major advantage of the contour crafting construction process is that it can print out any custom house design that can be envisioned. Professor Khoshnevi's work has already attracted the attention of NASA, with the idea that contour crafting technology can be adapted to build a colony on the Moon (Hopewell, 2012).

∞
The Amaze Project Brings 3D Printing into the Space Age

The European Space Agency has unveiled plans to bring 3D printing into the space age by using the new technology to build satellites, spacecraft and components for fusion reactors. The Amaze Project embodies the concept of "Additive Manufacturing Aiming towards Zero Waste and Efficient Production of High-Tech Metal Products." The project brings together 28 separate partners from European industry and academia, with factory sites being set up in the UK, France, Germany, Italy and Norway, with the objective of developing a viable supply chain.

Additive 3D technology has already revolutionized the design of plastics and has also been quietly integrated into automotive and aerospace industries. The Amaze Group has already begun printing metal jet engine parts and airplane wing sections up to two meters in size. Such high-strength components are normally constructed using exotic metals such as Titanium, Tantalum and Vanadium but use traditional casting, milling and grinding techniques, which tends to waste these expensive materials. With the new 3D additive manufacturing process, parts are constructed layer-by-layer from a digital computer model which produces nearly zero waste. Printing components as a single piece can eliminate welding, riveting and bolting which makes them stronger, lighter and less expensive in terms of skilled labor, time and cost. This new layering technology can also be used to create intricate designs, which would be impossible with conventional metal casting. This is important, since weight reduction of even 1 kilogram can save hundreds of thousands of dollars over an aircraft's lifespan.

According to David Jarvis, ESA's head of new materials and energy research, "We want to build the best quality metal products ever made. Objects you can't possibly manufacture any other way." He added, "Our ultimate goal is to print a satellite in a single piece, one chunk of metal that doesn't need to be welded or bolted. To do that would save 50 percent of the costs – millions of Euros." According to Jarvis there are still some problems that need to be overcome, such as with small air bubbles in the end product and rough surface finishing. "We need to understand these defects and eliminate them if we want to achieve industrial quality. And we need to make the process repeatable – and scale it up. We can't do all this unless we collaborate between industries – space, fusion and aeronautics (Morgan, 2013).

3D Printed Buildings for the Moon

Recently, architects Fosters and Partners announced that they have been actively engaged in designing 3D buildings for a moon colony that would be printed from regolith on the lunar surface. In their lunar scenario an inflatable structure containing a large robotic 3D printer would be transported from Earth to the lunar South Polar Region. The first structure to be constructed would be a four-person shelter. Other similar modules could then be created as needed.

In 2010, a scientific team from Washington State University determined that lunar regolith was indeed suitable for creating solid objects with the 3D printing process. The group formed a consortium with the European Space Agency, their primary objective being to test the practical feasibility of using a large 3D printer, similar to the one used in building housing units on Earth. A large vacuum chamber was used to simulate the lunar environment. According to Xavier Da Kestlier, partner in the Fosters firm, "As a practice, we are used to designing for extreme climates on Earth, and are exploiting the benefits of using local sustainable materials." With regard to the concept of bringing the 3D printing process into space, he stated, "It has been a fascinating and unique design process, which has been driven by the possibilities inherent in the material."

Another U.S. company, Deep Space Industries, has announced plans to harvest valuable minerals from asteroids. Their approach will utilize a 3D printer called the MicroGravity Foundry, which they are developing specifically for asteroid mining. The company plans to fully develop this new technology by 2020 (BBC News, 2013; Keane, 2013).

The Urbee Hybrid: The First 3D Printed Car

Two US companies, Stratasys and Kor Ecologic, teamed up to produce the world's first car to have its entire body printed by 3D additive printing technology. The new hybrid vehicle, named "Urbee" by its creators, has three wheels – two in the front and one in the rear. A prototype vehicle competed in the 2010 X-Prize Competition and got up to 100 mpg under city conditions and 200 mpg on the highway. The two companies have not yet revealed plans for mass production, but have high hopes for launching a revolution in the production of prototype vehicles. According to Jim Kor, president of Kor

Ecologic, "FDM (3D printing) lets us eliminate tooling, machining and handwork, and it brings incredible efficiency when a design change is needed. If you can go to a pilot run without any tooling, you have advantages" (Schwartz, 2010).

A Promising Technology for Transforming Air Pollution into Baking Soda

Coal-fired electrical generating plants produce carbon emissions, which pollute our atmospheric commons worldwide. The U.S., China and India alone planned to build 850 coal-fired electrical generation facilities by 2012. These new power plants will spew five times the amount of carbon dioxide into the atmosphere as the Kyoto Protocol nations had intended to eliminate. It is commonly thought that carbon sequestration, where carbon dioxide emissions are intentionally trapped and stored, may be the best solution for this global dilemma. To date, proposed solutions to this problem mainly involve storage in underground caverns, which is a very costly process.

A Texas-based company, Skyonic, has developed an innovative new technology, which employs a series of plastic mesh screens that can capture up to 90 percent of the carbon dioxide emissions produced by a coal burning power plant. The trapped carbon dioxide is then mixed with sodium hydroxide to produce sodium bicarbonate (baking soda). Since solids are relatively easy to store and transport, and since high-quality baking soda is the end product, this baking soda can be recycled for industrial use or even food ingredients. In 2006, a pilot plant for demonstrating the commercial feasibility for the new process was installed in the Texas utility, Luminant. Skyonics is presently continuing its efforts to upgrade this basic design to commercial scale (Obrien, 2010).

Bioplastics: An Eco-Sensible Strategy for Reducing Plastic Waste

Ecovative, a company in Green Island, New York, has come up with a new and innovative solution to the burgeoning problem of non-biodegradable plastic waste. Founders, Eben Bayer and Gavin McIntyre, are convinced they can create a line of mushroom-based products, which will take the place of most conventional plastics. By joining a group of other like-minded entrepreneurs they have committed themselves to making Planet Earth a greener and less toxic place. They expressed their primary objective in the following simple terms: "Our goal is to eliminate as many plastics as possible." Plastics have been considered an environmental nightmare for several decades. Since most plastics are petroleum-based they contain numerous carcinogens but tend *not* to break down in landfills. These plastics now clutter even the most remote beaches and land areas worldwide.

Five years ago the Ecovative team discovered that mushroom mycelia (the white thread-like root filaments) were able to bind particles like glue such that the resulting material could be used for insulation. In its ongoing quest for environmentally friendly plastic alternatives the company has already marketed it mushroom-based packing materials to companies like Steel Case, Crate and Barrel and Dell Computers. Revenues from these sales were expected to top $3 million in 2012.

At the Ecovative plant in upstate New York, "food" for the mushrooms is prepared in a clean room using organic waste products such as cotton hulls. Before the fungi can bloom into conventional mushrooms the mycelial threads are harvested for use in their products. Although Ecovative's products presently cost about ten percent more than non-biodegradable plastics the company feels it can lower its manufacturing costs once production is brought up to commercial scale.

Meanwhile, other start-up bioplastic companies are working to transform organic products such as chicken feathers, dried algae and soy into biodegradable bioplastics that will decompose naturally in landfills. Although at the time of this writing bioplastic firms compose less than one percent of the 560-billion-dollar plastics industry, the fledgling bioplastics industry is growing at a rate of some 20 percent a year. Ecovative feels they have a slight advantage over bioplastics made from agricultural crop waste or bacteria, as some waste is created in these processes. With mushrooms the threadlike mycelia actually "become the plastic" so the process is ecologically efficient. CEO Bayer predicts a bright future for his company and the bioplastics industry in general, stating: "We're where plastics was fifty years ago" (Machan, 2012).

In May of 2009 Coca Cola began using "plant bottles" composed of 30 percent bio-based polyethylene synthesized from sugar cane. In 2012, Pepsi Cola went a step further with plans to debut bottles made from 100 percent bioplastics. According to Melissa Hockstad, Vice President of Science, Technology and Regulatory Affairs for the plastics industry, industries are now increasingly focusing on sustainability. She states, "As the months go by I see more and more plant bottles on the store shelves, so it's definitely growing" (*Futurist*, 2012).

Converting Waste Plastics into Diesel Fuel

A London-based company called Cynar Technology has developed a new process for converting a wide variety of plastic waste products into diesel fuel. The Company claims to have created a profitable solution for the global waste recycling industry. This process solves a major issue facing the waste disposal industry since it also yields diesel fuel, which burns cleaner, and has a higher cetane rating and lower sulfur content than conventional diesel fuel. The Company has a patent pending for their new process and has established its first full-scale plant in Ireland. A second plant has been granted approval in the UK and is moving ahead according to plan. Cynar Technology is also moving forward with its sustainable waste solution that removes end-of-life plastics from landfills. The company is seeking new partners to become involved in this win-win technology for transforming waste plastics into clean diesel fuel. (www.cynarplc.com/index.asp)

Although technologies for converting waste plastics into fuel have been under development for decades, commercializing these technologies was previously considered prohibitive when the price of crude oil was relatively low. As costs for crude oil have risen, so have concerns about energy security and the environment. This has generated a renewed interest in the plastics-to-fuel recycling process. Scientists are hopeful the new technologies will ultimately provide oil-dependent nations with cheaper alternative fuels that can help reduce their foreign oil dependency.

To put this in simple perspective, a single plastic grocery bag can be converted into 10 milliliters of fuel. Thus, 100 grocery bags yield a liter of fuel that can power an automobile for five or six miles. Since the plastics industry is the third largest manufacturing sector in the United States, plastics touches nearly every aspect of our lives. Worldwide plastics pollution has become an increasingly serious problem in even the most remote locations of the world.

According to the Environmental Protection Agency, Americans produce about 31 million metric tons of waste plastics each year. However, only about 10 percent is recycled, since there are so many different kinds of plastics. Thus, plastic waste has become a major non-biodegradable component of municipal landfills.

According to Moinuddin Sarker, Vice President for R&D for Connecticut-based Natural State Research, Inc., a metric ton of plastic waste can yield eight to nine barrels of NSR fuel (a barrel being equivalent to 42 gallons). The company is developing a different process than Cynar. Plastic waste is first heated, causing the resulting slurry to vaporize. When the vapor cools, it condenses into a liquid, which, according to Sarker, "works in any internal combustion engine." Unlike conventional gasoline, which contains sulfur, nitrogen and phosphor, their final product, (NSR Fuel), is composed of only Carbon and Hydrogen. It is thus cleaner burning and less polluting than conventional fuels. The company has already demonstrated the process on an experimental scale, which has produced five gallons of fuel from a variety of plastics that have not been cleaned or sorted. Sarker posits that if the nation's annual plastic waste tonnage were converted to fuel, it would add up to 270 million barrels. To put this into perspective the U.S. consumes 21 million barrels of oil a day (Jean, 2010).

Liquid Air Technology:
A New Way to Store Intermittent Solar and Wind Energy

One of the major challenges for bringing renewable energy to the marketplace is the intermittent nature of wind, solar and tidal power generating stations. Conventional electrical power grids are created to provide peak demand power on a real-time basis. However, they lack the capability to store generated power for when it is most needed. A recent BBC News article showcased an emerging technology which could solve this problem, creating a new way to store intermittent renewable energy production by smoothing out the peaks and lows which are typical of these alternative energy sources. This would solve the main problem for these intermittent technologies by increasing their cost-effectiveness and practical feasibility. Most importantly it would make electrical power available when it is most needed.

According to environmental analyst Roger Harrabin, a new UK-based company, Highview Power Storage, is developing a process which uses waste electricity produced during low grid demand hours to create liquid nitrogen (liquid air), which is chilled and stored in high-pressure tanks. When peak electricity demand rises, the liquid air can be returned to room temperature. The expanding gas can then be used to spin turbines that generate electricity. As the nitrogen returns to normal temperatures it can be recaptured under pressure and the process begins all over again. Highview claims their system is only about 25% efficient by itself, but that by combining this technology with conventional power generating stations, their system can use waste heat from the station. Within this co-generation scenario Highview feels they can boost the efficiency of their process up to as much as 70%. The significance of this type of arrangement is that, in practice, power stations tend to hold their oldest, least-efficient facilities in reserve, using them only during periods of peak power demand. Research has shown that even modest use of renewable energy sources *can* have a major positive impact on lowering both peak power rates and exhaust pollution levels.

The new Highview liquid air system has been in operation for two years at a power generating station in Buckinghamshire. According to the Institution of Mechanical Engineers, "Liquid air can compete with batteries and hydrogen to store excess energy generated from renewables." They state, "The simplicity and elegance of the Highview process is appealing, especially since it addresses not just the problem of storage, but also the separate problem of waste industrial heat" (Harrabin, 2012).

Multiuse Titanium Dioxide: New Compound that Can Generate Hydrogen and Produce Clean Water While Processing Wastewater

A scientific team led by Professor Daren Sun of Singapore's Nanyang Technological University, has developed a unique new material they call "multi-use titanium dioxide." Turning ordinary titanium dioxide crystals into a patented microfiber, which can be easily made into flexible filter membranes, forms the compound. These membranes include different combinations of carbon, copper, zinc or tin – depending on the specific end product required. Titanium dioxide is cheap, abundant and has the ability to bond easily with water. It can thus function as a catalyst to accelerate specific chemical reactions.

Professor Sun's research team has created a new product which has the potential to provide efficient, cost-effective solutions for solving one the world's biggest environmental challenges – transforming waste materials into cheap renewable energy and clean water. The new titanium nanomaterial is claimed to achieve the following objectives: 1) When exposed to sunlight this nanomaterial can produce both hydrogen and clean water. 2) The material can be made into low-cost filtration membranes that are resistant to antifouling microorganism. 3) Nanofiber membranes can be fabricated into cheap, efficient membranes for use in reverse osmosis to produce clean water from saline or polluted water. 4) The membranes can be used to recover additional energy from desalination brine and wastewater. 5) The membranes can also be used to create low-cost flexible solar cells for generating electricity. 6) The life of conventional lithium ion batteries can be *doubled* when multiuse titanium dioxide is used for the anode. 7) The titanium nanofiber mesh can be made into bandages that are resistant to conventional pathogens.

In the five years it took to develop this new product Dr. Sun's research team discovered that multiuse titanium dioxide could also function as a "photo-catalyst" for turning wastewater into hydrogen, oxygen and clean drinkable water. This type of water-splitting is normally accomplished using platinum as a catalyst, but platinum is both expensive and rare when compared to titanium. With this new approach it is possible to treat wastewater and, at the same time, facilitate the storage of solar energy as hydrogen. Hydrogen is a "clean" fuel and can be used to power automotive fuel cells or power generating plants at any time of day or night – even when the sun is not shining. To date the Singapore research team claims to have achieved approximately three times the energy with their new multiuse titanium, as opposed to using titanium only as a catalyst. The icing on the cake is that the process can also produce clean water for nearly zero energy costs. Thus, this new process could offer the potential for upgrading water reclamation technology in urban areas worldwide.

Professor Sun has published over 70 scientific papers on multi-use titanium dioxide and its applications. He states: "While there is no single silver bullet with regard to solving the world's biggest challenges: cheap renewable energy and an abundant supply of clean water; our single multi-use membrane comes close, with its titanium dioxide nanoparticles being a key catalyst in discovering such solutions. With our unique nanoparticle material we hope to convert today's waste into tomorrow's resources, such as clean water and energy" (Nanyang Technical University, 2013).

Thorium Nuclear Reactors:
A Bridging Technology to Bypass Dangers of Conventional Nuclear Reactor Technology

Thorium has been called "the energy source of the future." The element thorium has several major advantages over uranium in terms of cost, energy efficiency and less environmental contamination. In a Liquid Fluoride Thorium Reactor (LFTR), thorium is dissolved in carrier salts to form a liquid fuel. This liquid salt fuel is pumped back and forth between a critical core and an external heat exchanger, where heat is transferred to a nonradioactive secondary salt that, in turn, transfers its heat to power a steam or closed-cycle gas turbine electrical generator. The technology, which was first investigated by the Oak Ridge National Laboratory in the 1960's, has recently attracted renewed interest worldwide.

The advantages of thorium salt-based reactors are as follows: 1) Thorium is far more abundant than uranium and is widely distributed throughout the world. 2) LFTR reactors produce only about 3% of the waste of conventional light water reactors. Due to the shorter radioactive decay period, waste products need only be stored for ten years and can then be re-processed to recover additional fuel – a far cry from the 300,000-year storage time required for conventional nuclear waste. Thorium liquid salt reactors can harness up to 98% of the energy of the fuel consumed, when compared to existing nuclear reactors, which operate at efficiencies of between two to five percent. 3) Molten salt in liquid form expands when heated, which slows the nuclear reaction and thus constitutes a *much safer* technology than with existing nuclear fuel-rod reactors. 4) Thorium salt-based reactors are essentially self-governing, which eliminates the possibility of meltdown – a major drawback of conventional fuel-rod reactors. Thorium reactors are designed with a plug at the bottom of the molten salt containment vessels. In case of emergency overheating the plug melts and the molten salt drains off into a shielded underground container. 5) Since thorium reactors *do not* produce plutonium, they significantly reduce the possibility of terrorist threats. 6) Molten salt reactors are not limited to using thorium fuel but *can* consume conventional nuclear fuel if necessary. This makes it feasible to dispose of the depleted uranium and plutonium produced as a waste product by conventional nuclear reactors.

Thorium-based reactors have already been built and operationally tested. From 1983 to 1989 a prototype thorium plant was built in Germany and three plants operated in the U.S. from the late 1960's to early 1980's. Although the technology performed according to plan, it was most likely abandoned because plutonium produced by conventional uranium reactors was then deemed "strategic" for military purposes during the cold-war era (Murray, Peter, 2012). Since Norway has major thorium deposits, the Norwegian government is presently working with U.S.-based Westinghouse and Norwegian-based Thor Energy to run a four-year test of thorium in a government-controlled reactor.

Other countries interested in developing LFTR technology include Japan, China, the UK and corporations in the US, Czechoslovakia and Australia. China has 14 conventional nuclear power generating plants already completed, with another 25 under construction. China also plans to build a thorium nuclear reactor and improve the existing technology. Although LFTR technology has its critics, there appear to be several major advantages to thorium-based reactors over conventional uranium reactors, especially with regard to cost, efficiency, security and environmental impact (Murray, Peter, 2012; op. cit.).

Precursor Engineering as an Aspect of Future-Science Energy Technologies

The term Precursor Engineering was introduced by pioneering physicist and engineer, Dr. Tom Bearden. The foundations for this technology were originally established by Paul Dirac, who proposed that we can "tickle" the quantum vacuum into containing not only positive and negative energy, but also negative energy and negative probabilities. Dirac even suggested that we have the potential to engineer physical reality itself! His hypothesis created such a backlash among his scientific colleagues of the 1920's and 30's that they discredited him as a scientist, then proceeded to eliminate any references to negative energy from standard physics theory – all because of the implications of his work.

Currently, leading physicists like Dr. Dan Solomon, Dean of the College of Mathematics at North Carolina State University, have published scientific papers highlighting the fact that arbitrarily removing negative energy from the basic equation violated the scientific method. Their papers clearly demonstrate that this action was in total error and thus violates the very foundations of science (Solomon, et al., 2004).

Victor Klimov of the Los Alamos National Laboratory has published research papers, which prove conclusively that real systems, *can* be constructed – systems which extract and utilize excess electromagnetic energy directly from the seething virtual-state vacuum. Laboratory experiments as well as replications in other laboratories have fully validated Dirac's work within the full requirements of mainstream scientific methodology (Bearden, 2009).

According to Tom Bearden, "The basic principles of precursor engineering open new possibilities for directly engineering physical reality itself." Within this new scientific framework "brute force" is not required to accomplish this. The field of precursor engineering could thus revolutionize conventional chemistry and physics by allowing us to tap into the quantum vacuum sea in a gentler way – a way which would be more in harmony with the laws of Nature. Bearden suggests that within this new scientific framework scientists can achieve beneficial results previously thought to be inconceivable. As an example he cites the environmental problems with hydrocarbon waste pools at the 1300-plus coal-burning power plants that presently exist in the U.S. He feels that new applications of precursor engineering technology might be applied to neutralize these carbon wastes, using Dirac's "tickling" methodology to transmute the polluting agents quickly and cheaply. He also suggests that this same technology might be used to *transmute* the massive accumulations of radioactive waste products from nuclear power plants the world over. This process would thus significantly shorten the hundreds or thousands of years associated with natural radioactive decay and render these waste products benign within a much shorter timeframe.

According to Tom the potential applications for precursor engineering can also be extended: 1) to eventually cure any disease quickly, cheaply and easily; 2) to draw cheap, clean electrical power directly from the vacuum; and 3) to convert desert areas of the world into inexpensive food-producing regions. These are just three examples of the potential benefits this new technology has to offer (Bearden, 2009, op. cit.).

MASTER KEY 12
Interspecies Communication and Interaction

> "We need another and a wiser and perhaps a more mystical concept of animals...
> In a world older and more complete than ours they move finished and complete,
> gifted with extensions of the senses we have lost or never attained,
> living by voices we shall never hear.
> They are not brethren, they are not underlings;
> they are other nations, caught with ourselves in the net of life and time,
> fellow prisoners of the splendor and travail of the earth."
>
> *Henry Beston*

Animal Communicators: Creating a Bridge Between Humans and Other Species

In many respects our animal companions are more like family members than "pets." The term "pet" refers to an animal that lives with humans for companionship or amusement. Today it seems that more and more people are realizing that these special animals are truly intelligent beings with their own life-purposes and missions. As such, they often demonstrate their own unique expressions of unconditional love to the humans and to other pets that have become part of their extended families.

Human-animal communication serves to create a bridge in understanding between humans and animals – whether they be pets or in the wild. Through our intuitive senses we can engage in meaningful dialogues with animals and learn to receive the subtle messages from those who share and co-exist with humans in the global environment. If we begin from a place of respect and reverence for all life we can indeed establish a special rapport with our pets and wild companions. We can learn from them, honor their space and thus learn to live more in harmony with the natural world.

During the past few years' professional animal communicators have quietly established a firm foothold in the fabric of society. Animal communicators are "pet psychics" who have developed the ability to communicate with animals and can thus serve as mediators to help resolve issues between animals and their owners. Animal communicator Nedda Wittels does not feel comfortable with the term "pet psychic." She feels that she is "not simply reading an animal as if it were a deck of tarot cards, but rather having an intelligent conversation with another intelligent being." She states, "Animals are sentient beings with feelings. They are not toys to be played with and discarded when inconvenient. They are alive and aware. They think. They feel. They make choices. When they are born, and for some length of time, they are similar to human children in that they require additional care and nurturance. Then, like human children, they grow up, albeit into a body, which may look small and cute, but which is adult. At this point, they are capable of making decisions for their own lives and should be treated with the same respect and honor that you would give a human equal" (Wittels, 2010).

It has been my own experience that animal communicators perform a valuable function in resolving issues between animals and their owners. This can take the form of simply functioning as an intermediary for animals to communicate with their owners and *visa versa*. Surprisingly, many emotional issues and behavioral problems (for both pets and their owners) can be resolved in this way. Another primary function of animal communicators is to assist with unresolved issues when a beloved animal dies. In many cases the passing of a pet can be as traumatic as the passing of a spouse, child or parent. Animal communicators can provide a valuable service by facilitating communication between a deceased pet and its owner, thus helping to clear up any unresolved emotional issues for both parties.

South African Animal Communicator, Anna Breytenbach

Anna Breytenbach is an animal communicator who has been practicing for 12 years in South Africa, Europe and the U.S.A. She works with both domestic and wild animals. Her "patients" have included: cheetahs, lions, wolves, baboons and elephants. She specializes in education and conservation-oriented programs. Anna's stated goal is to raise public awareness and to advance relationships between humans and other species – on both personal and spiritual levels. Anna's personal mission is simply to be a voice for animals and for the wilderness. To see Anna's amazing ability to interact with a black leopard, this fascinating documentary film can be found at: www.youtube.com/watch?v=wL--zc1KIxk. Anna's website can be found at: www.animalspirit.org

The Passing of My 19-Year-Old Cat Companion

I had a beloved male cat named Daisy who I trained to follow me like a dog as we took walks together on our creek-side property in Arizona. As Daisy got older I took our vet's advice and decided to keep him inside to protect him from wild predators like coyotes and bobcats that roamed the area. I built a small screened porch for him attached to the outside of the house, which he could access through a pet door. He could sit outside and still feel safe, while he enjoyed the natural surroundings. Later on when my former wife passed away, Daisy suddenly became a devoted companion, easing my grieving process and providing feline companionship for me. Daisy suffered from kidney degenerative disease, but he somehow managed to survive for two more years before passing away at the age of 19.

A week or so before my beloved companion passed, I contacted an animal communicator. She came to the house and sat with Daisy on her lap and communicated with him. At her suggestion I had purchased a dozen roses, which he told her he liked very much. She suggested I play my electric violin for Daisy, which he also told her he enjoyed. During the course of her dialogue with him, Daisy told her how it was a burden for him to continue to remain in his physical body, and now it was time for him to leave. He told her he was troubled about leaving me alone. Daisy told her that the reason he had pestered me, by reaching up and poking me with his paw when I was at the computer, was so I would be sure to take periodic breaks. He told her that he had known all about the pack rat that had gotten into the car I had parked outside, causing extensive damage to the wiring system. He also told her he would like to be buried under a nearby juniper tree, where he remembered spending many happy hours.

A few days later I woke up and heard Daisy gasping in his bed beside my own. I picked him up and took him into my office where I held him in my lap on a blanket. I soothed and comforted him while he took his last breaths and quietly passed away. Next morning I arose early, dug a grave for him under his favorite juniper tree, wrapped him in one of my former wife's skirts and sprinkled a bit of her ashes and some petals from the roses I purchased earlier. While this was truly a very sad event in both our lives, our session with the animal communicator helped me tremendously to process the grief of parting for both of us. In subsequent visits she would tell me when Daisy would come to visit in his light body. This brought much happiness and resolution for both Daisy and me.

For those interested in animal communicators I have included several books on animal communication in the Bibliography (Herzing, 2004; Randour, 2000; Roads, 1985; Robbins, 1997; Smith, 1999 and Wittels, 2010). A directory of animal communicators can be found at the following website: www.animaltalk.net/consultlist.htm

As a closing note on the subject of animal communications I have included links to the following two amazing videos which document how a Russian lady free diver, using meditation and special breath-holding exercises, overcame her fears and interacted with beluga whales beneath a hole cut in the ice – naked, with nothing to protect her from the cold. These videos depict a heroic example of how this lady was successful in surpassing the limits of the human body. www.youtube.com/watch?v=uuRJkrXL1_E and www.youtube.com/watch?v=oz4-2WgH47E

MASTER KEY 13
Subtle-Energy Management

> "Breakthrough involves creating a new pattern or paradigm,
> rather than improving on an old one.
> Improving the old is perfect change.
> Creating breakthrough is transformative change.
> There is a transformation to a new state."
>
> *Lindsay Collier – The Whack a Mole Theory*

When You Change the Energy You Change the Mass

I was once privileged to have a mentor who explained to me just how the above physics principle and its corollary can be applied to our daily lives. It works this way: whenever you become dissatisfied with a particular situation in your life you can *consciously* change the old energy patterns which led to your present situation and thus facilitate the shift into a more positive situation for yourself.

Changing the energy can be as simple as re-arranging your home. Another example might include taking a weekend to do something entirely new, travelling somewhere you have never been or attending a special movie or concert. You may surprise yourself at how this kind of unique experience can shift life energy flows into a more happy and positive framework.

By breaking old established patterns and bringing something fresh and new into our lives we discover that not only does our situation change, but we also begin to view the world from an entirely new perspective. By rearranging the furniture or getting a new hairstyle we are effectively changing the *mass-relationships*, which in turn cause the *energy relationships* in our homes and professional workspaces to shift correspondingly. By experimenting with this principle we will soon discover that applying this simple practice of energy management will serve to keep our lives adventurous, challenging and fulfilling.

∞

The Hundredth Monkey Effect as an Aspect of Quantum-Field Technology

The Japanese Monkey, *Macaca fuscata*, has been the subject of scientific investigations for over 30 years. In 1952 on the Japanese Island of Koshima, scientists had been providing specific monkeys with sweet potatoes that were dumped into the sand. A young female monkey named Imo discovered she could remove the gritty sand by simply washing her potatoes in the nearby ocean. Imo then taught this trick to her mother. Other monkeys in the group also learned this washing routine and the social innovation was passed on to some of the other monkeys in the group.

Over the next eight years *all* the young monkeys had learned to wash their sweet potatoes before eating them. Although the adult monkeys who imitated their children learned this new behavior, many of the older monkeys kept on eating the sandy sweet potatoes. Then, suddenly in the fall of 1958 a strange shift occurred. Although the exact number of monkeys who had learned to wash their food was not known, the theoretical hundredth monkey finally learned to wash its food. Apparently, the added consciousness of this hundredth monkey had managed to create the necessary "critical mass consciousness" to empower the quantum shift which followed. Suddenly, all the monkeys began to wash their food in the sea. If this astonishing event startled the scientists who were observing the monkeys, it was nothing compared to their surprise when they discovered that this new habit of food washing had somehow inexplicitly transferred itself to colonies of monkeys on nearby islands. This behavior was also taken up by a mainland troop near Takaskiyama.

This key shift in the social behavior of the monkeys was significant in that the mass of individual consciousness was apparently enough to shift the species consciousness into a quantum mode. "Although the exact number may vary, this Hundredth Monkey Phenomenon means that when only a limited number of individuals discover a 'new way' it remains the conscious property of those individuals. However, when one more individual manifests this new awareness, the field is strengthened incrementally. A critical mass is reached and the new awareness becomes the conscious property of all. This new awareness is apparently communicated mind-to-mind."(Keyes, 1982).

Financial Energy Management: Money as "Frozen Energy"

Since money plays such a significant role in our current global social perspective, I decided to include this brief section, which was taken from my first book, *Life Management 3000: Success and Survival in the Third Millennium*. Money, itself, is neither good nor bad. Accumulated wealth can be compared to water stored in a lake behind a hydroelectric dam. A certain amount (income) flows down from the mountains into the lake, and a certain controlled amount (expenditures) is allowed to flow through the hydroelectric turbines to generate energy that can be used in many ways.

In the game of life it is important to structure our financial energy so we are constantly building our assets, while keeping our financial outflows reasonably in line with income and savings. This approach to financial energy management insures that our lives will remain relatively stress-free, as opposed to those who constantly operate within an environment of extended credit and leveraged assets. Just stop for a moment and consider what a different world this would be if everyone followed these simple guidelines in their financial energy planning (Maynard, 2002).

The Physical Plane as a Training Ground: Learning to Control Energy and Observe the Results

The physical plane is a fascinating laboratory where we can practice managing energy and observing the results. This allows us time to adjust before we integrate with the higher planes of existence, where energy patterns manifest instantly and are thus harder to master. When we learn to raise our energy to come into alignment with the higher realms we discover that feelings and thought-forms can manifest almost instantly. The same is true of negative thoughtforms. Since we create the events we experience, the *quality* of the events in our daily lives is thus a direct reflection of our higher selves.

The more we focus and concentrate our energy the faster it vibrates and the more we express the creative universal force. When this occurs, dimensions that vibrate at a slightly higher frequency begin to open for us. This kind of "opening" is a key to a fulfilling life and a key to learning how to effectively manage the creative force. Anything we can do to improve ourselves assists our higher self in evolving. As we reach higher levels of refinement we can experience a new sense of excitement and deep gratification, as worlds we never knew existed are revealed to us. Thus, the more positive energy we can accept and process, the faster it will propel us forward and the more we will evolve.

Hidden seeds for unique creative expression exist in all of us. If these inherent talents are restricted by programmed limitations they may never have a chance to develop. If, however, we make a commitment to express our creative flow, our spiritual guides will have something to work with and we will quickly get the help and support we need to realize our full creative potential. The more we continue to move toward the light, the more power will be provided for our support. As this process speeds up we will discover we can upgrade our personal and professional lives in a matter of weeks – instead of years.

Managing Energy Fields and Creating Sacred Space through Visualization and Intent

The science of manifestation is no longer exclusively reserved for spiritual adepts. It is a part of the natural order of the universe. Indeed, we do it all the time. The key to manifesting is learning to be *aware* of the process. We need only to create the space for those events and things we wish to manifest. Although it's true we do not necessarily get what we always want in life we can get what we *should* have.

Visualization is a *mental* process, generated by the conscious mind envisioning a *spiritual* process. Visualization is governed by intuition, which is essentially a bridge to the super-conscious mind. Visualization can thus be a useful tool for bypassing the conscious mind so we can avoid the deep-rooted preconditions and personal issues that tend to derail our judgment. The inspirations and impressions we receive during the visualization process exist *outside* the "normal" parameters of conventional human thought. Since we are all truly interdimensional we have the ability to open ourselves to a conscious awareness of things that exist in the higher realms. Thus, we all have the inherent capability to "live" the vision – and continue to evolve it.

During this era of accelerated change we continue receiving downloads from our spirit guides in the higher dimensions. Our human bodies can be regarded as "temples of light." As such, they are being packeted with encoded light energies that can be grounded down into formats which function efficiently in the physical realm. This process highlights our roles as "transformers" for the greater social consciousness. During this process we too become transformed. From this perspective we function as "capacitors," such that each of us holds different frequency sets for personal and global transformation. In this way we facilitate the shift in human consciousness from the third into the higher dimensions. This is accomplished by referencing from the higher sources and *living into* the continuously expanding reality.

During this transformation process we interact with our spiritual guides and mentors who operate from the seventh through ninth dimensions. In the downloading process golden light is infused with red light, so the higher frequencies can be grounded down into our human blood crystals. Once these new gridworks become firmly established, this new protocol can be spread through our family, friends and professional associates. By working together synergistically in the quantum field we can thus discover new ways of perceiving the world and each other.

Our consciousness tends to center around our life commitments, much as the planets revolve around the sun. These life commitments exist to guide us so we can organize our lives around those things and situations that raise our consciousness to higher levels. We tend to resist new commitments for fear we might be overwhelmed by the sheer magnitude of the promises we make. In doing so we sometimes lose our way in the process. This is why mainstream society tends to be so fixated on the dramas and distractions of a dysfunctional and commitment-free life style, since living in denial is "the easy way out." Without a firm set of life commitments we tend to become isolated and alone in life. Thus, interconnectedness and belonging relate directly to our willingness to commit – to one another and to those ideals and aspirations, which are the basis for our existence.

If you do not yet have a mission partner in your life, then commit yourself to an idea, cause or new pathway. This forms a matrix into which you can direct your creative energies. It is important to remember that many people exist in a world of pain, simply because they have nothing to care about, nothing to surrender themselves to and nothing or no one to love. Thus, there are millions of folks who need our love each day – millions of children, the elderly and animals who are simply withering away from a lack of love and any sense of belonging. The less we are willing to be responsible for the way our actions affect others around us, the more disheartened we will ultimately become with our lives. Thus, our ability and willingness to commit to the well-being of other individuals is predicated on our ability to commit to our *own* well-being and happiness.

Without first making a commitment to ourselves we become limited in what we can offer others. Thus, anything meaningful we might be able to offer anyone else in our lifetime begins with commitments we make to ourselves. This deep sense of personal commitment empowers our goals, visions and dreams. It is also the source from which true miracles are created.

For most of us, before we commit to something, we like to have proof that things are going to work out the way we want them to. In other words we all like a "sure bet." The catch here is that commitment must come *first* for manifestation to occur. Thus, we soon discover a unique synchronicity that occurs when we make a wholehearted commitment to ourselves or to a special goal or vision. When we have things lined up correctly this kind of "magic" appears to occur spontaneously – and often when we least expect it. Within this same context the following quote by German philosopher, Goethe, makes perfect sense: "The moment one definitely commits oneself, then Providence moves too. All sorts of things occur to help one that would never otherwise have occurred. A whole stream of events issues from the decision, raising in one's favor all manner of unforeseen incidents and meetings and material assistance, which no man could have dreamed would have come his way."

From a slightly different perspective it is important to understand that if we do not first honor *ourselves* we tend to attract people who do not honor us. Thus, it is important for each of us to develop our own strengths and character before we can to develop endearing and enduring relationships with others. Developing a strong sense of *self-love* and *self-respect* keeps us from losing our identity in our relationships with others. During this relationship-building process we need to be actively engaged in self-discovery at all times. After all, this is what evolution is all about.

In this process of self-discovery it is important to set reasonable boundaries – boundaries for caring for ourselves and for others. Establishing sensible boundaries is thus a key for achieving healthy and balanced relationships. Energetically speaking, well-established boundaries allow us to contain and share our energy in more balanced ways.

Shifting Frequencies, Shifting Fields

We exist in a "sea of frequencies." The shifts that accompany them can be extraordinarily beneficial and uplifting. Each time a frequency shift occurs our energy fields are correspondingly modified. Thus, we need to adjust our perspectives and strategies accordingly. Adjusting to each new shift often causes our physical, emotional or mental bodies to act up in response to being moved from their former comfort zones. Be prepared to take the necessary steps to correct these imbalances.

On the downside, electronic smog from cell phone towers, smart meters, wi-fi routers, cell phones, cordless phones and similar electronic devices can cause unwanted frequency shifts. For this reason, an awareness of these technical sources of interference is the first step in dealing with them. It is thus important to learn to shift frequencies *consciously*, as this is a key survival tool for sustaining our health and emotional balance. It is also important to understand that *frequency awareness* and *conscious frequency shifting* are evolutionary imperatives for humans and for the health of all living things.

Spiritual Responsibility and Guidance from the Higher Realms

Being open to information from the higher realms insures we can continue existing in a field of positive energy. True spiritual responsibility involves taking responsibility for *all* aspects of our lives, no matter what problems we encounter or who caused them. Being spiritually responsible is the opposite of taking on a "victim identity."

Spiritual responsibility is directly related to the strength of the connection between our crown chakras and the higher realms. A person with a closed crown chakra tends to have a "polite indifference" to God or the higher realms. Typically, this type of person prays when things go wrong, often trying to bargain or negotiate with God until things return to normal, after which prayer is usually put aside until the next time it is needed. Spiritual responsibility implies a strong and loving connection with higher sources. It should be free of imposed tribal beliefs such as fear, hardship and poverty. People who have this fear tend to be spiritually disconnected and their lives become empty and meaningless.

When we commit to spiritual guidance, our crown chakras remain open so we can receive guidance, joy and inspiration. Whereas the ego mind enjoys engaging in endless discussions and rationalizations, higher guidance is direct and enjoyable. If we require help we need only ask with sincerity and we will receive all the guidance and help we need.

During a recent ski trip to Purgatory in Durango, Colorado, while skiing down the mountainside, I began to hear a voice in my head, telling me to "turn here" or shift my weight slightly on a fast downhill run. I suddenly realized I had discovered a simple truth. If you ask for help and calm your mind, a teacher will appear to guide you. It's that simple. Just make the request and open your consciousness to receiving guidance.

MASTER KEY 14
AI/Human Energy-Field Interfacing

> "With the addition of artificial intelligences to the network, and the developing richness and quality of the flowing information, Marshall McLuhan's 'global village' has changed into the 'global brain' – an entirely new kind of planetary species."
> *Yatri, 1988 – Unknown Man: The Mysterious Birth of a New Species*

Is the Internet Becoming A Conscious Entity?

Brain scientist Jeff Stibel feels that the internet is essentially a "new life form" that is showing the first signs of intelligence. In a BBC News article he maintains that the physical wiring of the internet is similar to a rudimentary human brain, and that some of the actions and interactions that take place there are similar to the processes scientists have observed in the human brain. He claims that the internet is forcing humans to think and interact in new and different ways – and that this is just the beginning. "The internet is only going to become more and more intelligent, changing society in ways which we are not yet able to understand" (BBC News, 2012b).

This is also true of computer programs and the programming on our tablets and mobile phones. These digital devices all have their own basic levels of intelligence and are able to interact with their human counterparts, give suggestions and offer search alternatives. The best software programs, apps and widgets are "intuitive" and "user friendly." Two examples include suggestions by Amazon.com based on your past purchases, or reminders from E-Bay that an item you have bid on has been relisted.

Digital Prayer Wheels and Torsionic Meditation

In the 1970's a U.S. Air Force Group in Thailand donated some obsolete computer equipment to a local group of Buddhist monks who began experimenting with the hard drives. Their idea was to use them as modern techno-analogs of traditional Buddhist prayer wheels. Following the introduction of Tibetan Buddhism into the West many new types of prayer wheels have been created. Among these innovations include mechanical electric prayer wheels, solar prayer wheels and digital prayer wheels that can be downloaded to a computer or cell phone. Interestingly, his Holiness the Dalai Lama, has apparently given his approval to these technological innovations by suggesting that having a digital mantra on computer works as well as a traditional prayer wheel. Since a computer hard disk spins hundreds of thousands of times per hour, and can contain many copies of a mantra, anyone can now add a digital prayer wheel to their computer, tablet or mobile phone. (www.dharma-haven.org/tibetan/prayer-wheel.htm)

According to leading-edge consciousness researchers in Russia the groundwork for Torsionic Meditation was established by spiritual teachers Nicholas and Helen Roerich. From a Future-Science Technology perspective Buddhist prayer wheels and computer hard drives can be considered to be "generators" for propagating faster-than-light torsionic emissions. Such devices are thought to be capable of carrying messages even over interstellar distances.

Computer hard drives are essentially finely-tuned, spinning fractal multi-magnets. Since they are mass-produced, they offer the advantages of being relatively inexpensive, yet are built within precise manufacturing specifications. With millions of PC's in the world today, the possibility exists for creating a *Global-Scale Phased Torsionic Antenna* that would be effective for facilitating collective meditations or shifting the global consciousness.

Unique strategies for consciousness upliftment, peace and planetary defense (The Spiritual Defense Initiative) were originally suggested by spiritual teacher Swami Satchidananda, who influenced the formation of the Pentagon Meditation Club in Washington, DC and the Russian Initiative Group for Defense of Earth, whose members investigate the scientific exploration of "Inner Space" (Ivanenko, 2003). This subject is treated in greater detail in the section in Core Paradigm 25 (Quantum Leapfrogging).

The "Force-Magnifier Effect:"
An Effective Tool for Social Transformation

The "force-magnifier effect" involves the use of the internet, websites, blog sites, international television, radio talk shows and social media. A transformative concept is first created in visual or audio format. This "information meme" is then broadcast out to the world via social and global media networks. Through the various electronic, digital and consciousness modalities the impact of the original message is multiplied exponentially as it spreads throughout the global datasphere. As mentioned previously, the internet can be regarded as a "living entity" since it has the innate capabilities to grow and evolve according to the inputs and technologies associated with it. The internet thus represents a powerful "force-magnifier" which has the capabilities to amplify individual or group powers and thus influence others by putting new concepts out into the "global mindfield."

Welcome to the Future of Medicine! Star Trek Technology has Arrived!

Is it possible that a simple hand-held electronic device can accomplish miracle cures? You bet it can. It already exists! The Russian SCENAR hand-held electronic healing device is about the size of your remote TV controller and might be called the best miracle-healing device ever invented.

SCENAR stands for "self-controlled energo-neuro-adaptive-regulation." The device represents a brilliant fusion of western technology and eastern energetic healing methodology (virtual medicine). The origins of the device are shrouded in Russian military secrecy, but we do know that the original SCENAR device was invented in 1973 by electronics engineer, A.A. Karasev, who developed the new technology after some of his family members had sickened and died – despite conventional medical procedures then available.

When the Soviet Union began sending cosmonauts into space for prolonged periods of time, it became clear that a new approach to treating illnesses was a high priority. Unlike America's NASA Program with its reusable space shuttle, there was no convenient way for the Russians to evacuate sick cosmonauts back to earth for treatment. Conventional pharmaceuticals were not an option, due to weight restrictions and the fact that drug-related treatments are based on the principle that one drug is normally used to treat each specific condition. Thus, even a minimal medicine assortment would add unwanted weight. Also, in space-capsule environments, where water recycling is essential, any drug entering the water system would remain there and pass through the cosmonaut many times. It was thus decided that a radically new approach to this problem was needed. Any new device had to be easy to use, lightweight and medically effective. Since the field of bioenergetics technology was the only existing paradigm, which seemed capable of delivering this kind of new technology, the SCENAR concept came into the forefront. Ironically, due to funding delays and the establishment of joint ventures between Russia and the NASA programs, the Russians gained the capability to ship sick cosmonauts back to Earth on the space shuttle. Consequently, the development of the SCENAR device was abandoned and it was never actually used in their space program.

In the years that followed some of the original researchers went on and continued to develop the device for use in the global medical community. The most recent version of the SCENAR device is a hand-held unit, which resembles a TV remote control, weighs just over ten ounces and is powered by a standard 9v battery. During a treatment, the metal prongs on the device are placed in contact with the patient's skin and gently run back and forth over the affected areas of the body. Following the initial contact with the patient, the unit begins collecting electromagnetic signals from the body. These signals are modulated by software programs in the device, and then re-emitted back to the affected tissues and organs. Essentially, the device begins to use the patient's endogenous signals within a cybernetic feedback loop. It does this by scanning and re-transmitting the signals many times a second. What is remarkable about the SCENAR approach is that the unit apparently enters into a sort of dialogue between the artificial intelligence of the device and human body – especially the diseased organs or tissues. The device then evolves a new signal pattern for the diseased areas and shifts them into a healing modality. New frequencies and energy patterns are thus established, which in turn become fresh input signals to be further modified, and so on. This output-equals-new-input sequence is similar to the way in which fractal patterns are generated (Scott-Mumby, 2014).

The SCENAR treatment is a safe, non-toxic, non-invasive natural bioenergetic healing therapy. The unit is approved as a handheld therapeutic medical device for use in pain management. It essentially stimulates the brain and nervous system to restore affected communication lines within the body and prompts it to "re-start" the healing process. The device has been authorized for clinical use by the Health Ministry of Russia since 1986. It has now become a mainstream medical protocol in that county and has also gained increasing acceptance by health practitioners throughout the world.

SCENAR begins its ongoing repair work by stimulating the release of peptides from the body's "internal pharmacy." Peptides are a key to the healing process, since they regulate a multitude of physiological processes at the cellular level. Since neuropeptides are natural pain relievers, the release of these regulator chemicals often provides direct pain relief during treatments.

In addition, the body is also stimulated to release its natural "feel-good hormones," which is why so many patients comment on how calm and relaxed they feel after being treated with the device.

It is important to understand how the SCENAR device differs from other electrotherapy instruments such as TENS devices which are mainly used for pain relief, or electro-acupuncture devices which have a broader range of applications. The main difference is that both these other healing technologies lack the biofeedback capabilities of SCENAR, which is what makes it unique. Basic information, videos and patient testimonials on SCENAR devices can be found at: (SCENAR-Cosmodic, 2014; Beachley, 2012; Pinterest, 2014).

Electronics as Spiritual Wave Guides

With the advent of ultra-high definition video cameras and LCD displays, 3DTV, wireless routers and the proliferation of expanded broadband capabilities we often forget that information from the higher realms comes into the physical realm via light-encoded packets. The capability of the human brain to handle, prioritize and access huge amounts of data has been stimulated to expand correspondingly. Keeping up with this burgeoning information blitz is easier if we simply flick that switch in our minds to become aware of the fact that these electronic devices function as "consciousness wave guides." When we become conscious of this principle we open ourselves to receiving this information more easily. Next time you listen to a favorite piece of music or watch your favorite sitcom on HDTV, hold this thought in mind and you may be surprised at how this shifts your perception of the information being conveyed.

MASTER KEY 15
Future-Science Medicine

> "The human body, like everything else in the cosmos,
> is constantly being made anew every second.
> Although your senses report that you inhabit a solid body in time and space,
> this is only the most superficial layer of reality.
> Your body is something far more miraculous –
> a flowing organism empowered by millions of years of intelligence.
> This intelligence is dedicated
> to overseeing the constant change that takes place inside you.
> Every cell is a miniature terminal connected to the cosmic computer."
> *Deepak Chopra, 1993 - Ageless Body, Timeless Mind:*
> *The Quantum Alternative to Growing Old*

Future-Science Medicine involves merging state-of-the-art conventional medical technology with psychic abilities and consciousness technologies such as medical intuition, diagnosis and healing. The integration of traditional medicine with complementary and indigenous medicine creates a synergistic relationship between hard-wired technology and consciousness technology. This melding forms a synergy between the two medical protocols and thus offers the best of both worlds.

In my own past association with conventional medicine, I have been fortunate to be able to work with medical doctors who were associated with major alternative medical associations, as well as purely conventional surgeons and medical specialists. I have noticed that what outstanding doctors and scientists have in common is that they bring the *art* back into medicine and science. Despite the fact that they have come through conventional educational systems, most of these brilliant medical and scientific leaders have told me they acknowledge and rely on their intuitive senses – although most would not wish this to be public knowledge.

A Tiny Fish that Could Transform Biology and Medicine

At Duke University, geneticist Nico Katsanis studies the causes of rare illness. He is among the increasing numbers of researchers who have begun using zebrafish instead of lab rats in their research. These inch-long fish have recently come into favor over traditional lab rats and mice for three main reasons: 1) Zebrafish reproduce much faster than rats or mice. Although lab rodents take about three weeks to produce about ten offspring, a female zebrafish can produce hundreds of embryos only three days after spawning. 2) Zebrafish are much less expensive to maintain. It costs only about six and a half cents a day to maintain a tank of several dozen fish, compared with 90 cents a day to keep five rats or mice in a cage. 3) Because zebrafish are transparent, scientists can observe their organs and even watch them grow. This makes them ideal for research associated with organ regeneration.

In 1988 researchers had already learned to selectively mutate zebrafish to the point where they became acceptable models for studying human diseases. Since that time research papers on zebrafish had increased from a mere 26 to 2,100 by 2012. Due to increased demand for the new experimental fish a special non-profit Zebrafish International Research Center was organized. The center now produces 2,608 different genetically modified strains for research purposes and provides specially modified fish to nearly a thousand different corporations and research labs.

According to Harvard Medical School's Leonard Zon, "The field is on fire." Zon's lab uses fish models for his skin cancer, blood disease and stem cell research programs. Other labs have created fish with specific DNA mutations linked to human narcolepsy, muscle disorders and large head size associated with autism and rare diseases in children.

Zebrafish have also proved to be ideal test subjects for the initial screening of promising drugs prior to testing them on mammals. Compounds to be tested are simply added to the water containing the fish, and the fish absorb it directly through their skin and gill tissues. Zon's lab at Harvard was the first in the world to develop a new drug, which was discovered with this technique. Through this approach, lab workers were able to try 2,500 different molecules within just four months to discover one molecule that dramatically increased blood stem cell counts in the fish. Results from these initial tests were then validated using mice. Human clinical trials were begun in 2009 with 12 leukemia patients whose red blood cells had been destroyed from chemotherapy. The new drug boosted red cell counts rapidly in ten of the twelve subjects. Since that time Zon's group has used zebrafish screening to search for a drug to cure skin melanomas. At the time of this writing, initial trials of this drug on human subjects were well underway.

The advent of zebrafish test screening would thus appear to open the door for a wealth of new medical research opportunities since this new experimental technique dramatically reduces research time and costs for discovering new drugs and for finding cures for rare diseases (Hughes, 2013).

The Role of Medical Intuitives in Future-Science Medicine

Medical intuitives are psychically sensitive individuals who use their psychic senses to "see" the energy patterns in the physical and higher vibrational bodies of a patient. A medical intuitive is able to scan a patient's energy field to determine the location of infections, problems or blocks in their normal energy flows. Since disease conditions tend to show up as energy blockages *before* they invade the physical body medical intuitives can also identify vulnerable points in a patient's body, which could lead to future problems.

Some medical intuitives are also healers, but their modalities of perception and healing often differ markedly. Interestingly, there are a number of well-established medical intuitives who are already quietly working hand-in-hand with conventional doctors, surgeons and specialists. This practice is the extension of a historical tradition, as kings and nobility of the past often held advice from medical intuitives in high regard.

Among the most famous medical intuitives is Edgar Cayce, who was known as "the sleeping prophet," since most of his readings came from transcriptions of medical diagnoses which he received in a trancelike state. In the course of Cayce's career over 30,000 of his health readings were recorded by his followers. Subsequent research on these readings and patient histories demonstrate a direct correlation between these readings and follow-up patient diagnoses and cures. More on Edgar Cayce's works and the Cayce Foundation and can be found at: www.edgarcayce.org

Each different disease has its own "energy signature." Medical intuitives can often identify the severity of a patient's condition, and determine whether a given condition is life-threatening or not. This ability allows for skilled medical intuitives to work together with medical doctors and emergency room nurses to prioritize patients, based on their most immediate needs. Skilled medical intuitives can also be invaluable members of healthcare teams, since they can diagnose problems often overlooked by conventional medical practitioners.

Medical intuitives work by scanning the electromagnetic field that surrounds the human body – similar to conventional MRI's or other medical devices. This energy field reflects changes in the health of the physical body in accordance with the patient's state of health and vitality. In healthy people this energy field vibrates at a high frequency; when a disease condition is present it vibrates at a lower frequency, such that dark spots (as "seen" by medical intuitives) indicate blockages in the natural healthy flow.

The human energy field has two major components. The first is the *electromagnetic field*, which is located just outside the physical body. The second is comprised of seven energy centers or chakras. These seven chakras include the *root chakra*, located at the base of the spine, and the *crown chakra*, which is located above the top of the head. These chakras correspond to clusters of nerve cells in the body, called ganglia, which branch out from the spinal cord.

Medical intuitives base their diagnoses on the following principles: 1) Illness begins with a disturbance in the electromagnetic energy systems of the human body. 2) This energy is manifested as an electromagnetic field that surrounds the physical body. 3) The frequency of this vibrational field increases or decreases according to a person's general health and well-being. 4) All diseases are characterized by a lower vibrational frequency. 5) Since "Like attracts like," if our vibrational energy remains at a consistently low level it will tend to resonate with illness. Thus medical intuition represents an important key to diagnosing disease and developing new and unprecedented preventative medical procedures (Nani, 2004). Additional information and listings for medical intuitives can be found at the International Association of Medical Intuitives' website at: www.medicalintuitives.net

Personal Experiences in Intuitive and Modern Medicine

During my graduate school days I had the opportunity to spend time with a relative who was married to a famous children's heart surgeon in Philadelphia. While taking a brief holiday from my scholastic routine I stayed with the family and was invited to sit in the gallery above the operating theater at the Philadelphia Children's Hospital to observe open heart valve repair surgery on a seven-year-old child. Via intercom, I was able to communicate with the surgical team during this amazing experience. The surgeon later confided in me he often relied on his intuitive senses in his work. He also told me it helped him select the best people to train for his surgical team and that he was so closely connected with his patients at the hospital that he would intuitively know when to go back to the hospital to check on certain patients even before the hospital called him.

I have experienced the benefits of psychic and intuitive healings myself. At age 55, I had a severe auto accident where I fractured 13 ribs, a collarbone on my left side and a shoulder bone on the right. I was flown by helicopter to the hospital in the next town, and for the first 48 hours I had to consciously draw each breath due to the pain from the broken ribs and associated physical trauma.

As luck would have it, my supervisory doctor was a 75-year-old thoracic surgeon who relied on his sense of intuition in his distinguished medical career. I know this because I would catch him peering periodically through the doorway of my hospital room to observe my breathing, skin color and general behavior.

When I was released from the hospital, I called my long-time Hopi sister and medicine lady, Theodora Sockyma, who I had known for many years. She drove down from the reservation with her husband and spent about an hour doing healing work on me. She simply had me sit outside on a picnic bench and worked over my rib cage with a gentle massaging motion of her hands. After a couple of months of physical rehabilitation I was able to eventually resume scuba diving once again. [Since I had worked directly for the Hopi Tribe several years prior to this accident, I was well aware that this medicine lady had been officially sanctioned by the Hopi tribe to screen and approve any Native American healers who came to the reservation. She told me she had a good rapport with the Reservation medical doctors and that they had often validated her traditional healing methods using conventional x-ray technology].

A couple of years later I had another medical crisis which also involved intuitive medical advice and the same Hopi medicine lady, Theodora. At that time I suffered from debilitating headaches, but had not consulted a medical doctor. Instead I called Theodora and asked her if she would look into it for me. The next day I was outside working in my yard, when my wife told me that she was calling and wanted to speak to me. I took the phone and Theodora said, "Elliott, you better get yourself over to the hospital and have that gall bladder taken care of." Since she was uncharacteristically outspoken, I made an appointment with a surgeon at the nearby hospital who scheduled me in for a gall bladder scan. I then visited the surgeon's office to get his professional opinion. He said I did have a gallstone but the problem might simply go away with time. He did not seem particularly interested in performing surgery. Since I was planning a trip to Canada in a couple of weeks and needed the problem resolved quickly, I decided to take an intuitive leap! I said, "Well, Doc, I don't know exactly how to put this, but I have been close friends with a Hopi medicine lady for many years and she told me to get my butt over to the hospital and get this gall bladder problem taken care of. Can you schedule me in for surgery ASAP?" The doctor showed no visible reaction. He said, "OK. How about next Tuesday?" Three days later I was prepped for arthroscopic surgery and wheeled into the operating theater. As I was coming out of the anesthesia in the recovery room, the surgeon suddenly appeared. He was holding a plastic container with a gallstone the size of an olive – and a Cheshire cat grin on his face. The problem was thus resolved. I recovered within a few days and was able to travel to Canada as planned.

Over the years I have been blessed to work with many medical doctors, surgeons and alternative medical experts, each of whom had their own interesting stories to share with me. Complementary medical practitioners often use a technique they refer to as "kinesiology" or "muscle testing" (a simple dowsing technique) to help diagnose and treat patients. Ultimately, the most important objective is to achieve a positive medical outcome – whether through conventional or alternative medical practice. Over time, documented results can be used to scientifically validate complementary diagnosis and healing methods so they can be more easily integrated into a new future-medicine format.

Medical Intuitive and Healer, Jerry Wills

Jerry Wills is a skilled healer and medical intuitive who has been featured on numerous national TV and radio talk shows. He is also an experienced teacher and has presented a healing and energy workshops throughout the US and abroad. Among his many other talents he is an accomplished musician, electronics engineer, videographer and archaeologist. With his wife, Kathy, he has led several expeditions into the high Andes Mountains in Peru and Bolivia to explore ancient archaeological ruins.

At the insistence of his wife, Kathy, Jerry began his healing practice in Arizona over five years ago – [Kathy is also an accomplished healer]. Since that time he has shared his unique gift of healing with countless individuals – including indigenous peoples in very remote areas of the world. Although Jerry is the first to admit his healing is not always 100 percent effective, a high percentage of the people he has worked on have experienced immediate healing relief – often after years of pain and suffering. Curiously skeptical audiences typically attend Jerry's lectures and healing workshops. Jerry first introduces himself then asks anyone in pain to raise their hands and step forward to briefly explain their problem. After the first volunteers step up and experience positive results the line keeps getting longer so his workshops often last into the morning hours of the following day. Over the past year I have been privileged to spend quality time with Jerry and include him and Kathy to be among my closest friends.

I first learned of Jerry and his remarkable healing abilities nearly two years ago, when I was suffering from a painful bout of sciatica in my right hip and leg. I was unable to straighten up and experienced chronic pain most of the day and night. This condition drained my physical energy and negatively affected my formerly positive state of mind. At my wife's suggestion I called Jerry and subsequently arranged for a healing session with him at our home. When he and Kathy arrived I was impressed. Jerry is nearly seven feet tall and has a countenance and demeanor that is both charismatic and disarming.

Prior to this time, I had been treated by different types of alternative healers – all to no avail, so I was understandably a bit skeptical. Jerry placed his hands gently on my lower back and closed his eyes for a few moments. He psychically "scanned" my lower back vertebrae where a major nerve to my right leg was apparently pinched. He told me that he detected some inflammation but said, "It would not be difficult to fix." He then began the healing process, which took less than five minutes. He explained how he was able to follow the timeline back into the past to a time when my back had been healthy and normal. He then superimposed that time-point on the present condition to accomplish the healing. He explained how he used the body's own materials to repair damage such as torn cartilage or inflamed intervertebral disks. Jerry used the term, "reorganization of matter" to describe his methodology. He suggested I remain quiet for the rest of the day but told me he had corrected most of the problem.

The following day, after a good night's sleep and a few lingering doubts, I was surprised to discover that I could straighten up once again and was able to get outside and work with my tractor. I was delighted to find myself nearly free of pain. About two weeks later I arranged for a second session with Jerry and subsequently realized that I had been completely healed from a painful and debilitating condition. At about this same time, my wife, Alisa, although somewhat skeptical, arranged for a healing session with Jerry. She was experiencing severe pain from an earlier knee injury. Following her session with Jerry she was able to stop taking pain medication and resume hiking once again.

During one of my initial conversations with Jerry, I asked, "Where are you from?" He quietly told me how he was "found" in an abandoned Kentucky farmhouse which was located on a military base. He told me that a team of military personnel was there to pick him up as they had apparently been notified of his arrival. He went on to tell me how he was placed with foster parents. He did not discover the truth about his mysterious past until he was in his 20's when he pieced together the story from his foster parents and others. He said he was born in the Tau Ceti Star System and that the military personnel had apparently been informed that he and several other individuals like himself, were "seeded" on Earth to assist humanity in their evolutionary process.

I have personally spoken at length with two different individuals (Robert Dean and Charles Hall) who discussed in detail their formerly classified experiences in the U.S. military, which involved regular contact and dialogue with extraterrestrial beings. Based on my personal observations of Jerry's remarkable healing technique I am convinced he is telling the truth. Later on I discovered that Jerry has other incredible insights and abilities to offer. He is an accomplished electric bass player and performed in a rock group in his younger days. In addition to his extensive knowledge of human anatomy, physiology and advanced nutrition he has honed his skills to "heal" electronic circuits and devices such as sealed recording tape heads – a skill he has demonstrated to others in the past.

Jerry's healing abilities opened the way for his acceptance by isolated native tribes in Peru and elsewhere who would often disclose information about ancient ruins and sacred sites which had been kept hidden from outside archaeologists and explorers. Jerry and his wife have organized several expeditions to remote areas of Peru and Bolivia where they have explored and videotaped previously undiscovered archaeological ruins and artifacts.

Jerry is committed to sharing his rare gifts with other less fortunate individuals. He displays an amazing amount of humility and objectivity as he describes his adventures. I regard him to be one of the kindest most caring individuals I have ever met. Although Jerry charges for his healing sessions and workshops he has also healed many people who could not afford to pay. He has been more than willing to share his gifts with others by holding healing workshops where he teaches his students to heal others and themselves. You can watch a Fox News video clip of one of Jerry's amazing healings at: www.youtube.com/watch?v=J0e7uUPtALs. He is also featured in television interviews with Canadian exopolitics expert, Alfred Webre, at: www.youtube.com/watch?v=rI8a4oSJ6do. More on Jerry's healing workshops, contact information and ancient archaeology tours can be found at his websites at: www.jerrywills.com and www.xpeditions.tv/website-disclaimer.html

Future-Science Healing Protocols Based on Intuitive Medicine

The process of healing, as seen from a medical intuitive perspective is simple: The first step is to realize that some of the things our unconscious mind has been programmed to believe about ourselves are simply *not* true. This involves taking an "objective observer" position and looking to see just *where* we have been investing our life energies. By eliminating our negative energy investments and focusing on those which bring a positive energy return we can shift into a healing mode, based on feeling better about ourselves and operating within a self-generating state of happiness and inner peace.

The second step involves *claiming ownership* of any dysfunctional modes in which we have been functioning – most often through subconscious programming imposed by our family and social patterning. By recognizing old behavioral patterns as they surface we can correct and replace them with the more positive values of self-appreciation and happiness.

The third step involves healing any self-sabotaging beliefs – replacing them with positive beliefs, which will free us to work within a happy, healthy and creative framework. Since toxic beliefs tend to remain hidden in the subconscious, uncovering and acknowledging them, then working to replace them with more positive-energy beliefs, will allow us to *consciously* reinvent ourselves and move onward in a series of quantum leaps.

Curing Cancer, Radiation Poisoning and Other Medical Conditions with Cannabis: A Surprising Natural Medicine

Nuclear radiation has been released in our atmosphere, oceans, drinking water and food supplies Radiation has been released by nuclear devices detonated since 1945 and by the Chernobyl disaster in 1986. More recently, the earthquake and subsequent nuclear reactor meltdowns in Japan have effectively exposed all life in the Global Biosphere to increasingly higher doses of radiation. Since this invisible threat has grown slowly and its effects last for relatively long time periods it represents a "ticking time bomb." To date little progress has been made by medical science to reverse and heal the effects of this insidious legacy. In the face of this increasing nuclear pollution of the global commons the good news is: There may be a simple solution.

Rick Simpson of Nova Scotia, Canada claims to have developed a cure for radiation sickness, based on the use of hemp oil to cure patients who experienced severe side effects from undergoing chemotherapy as prescribed by conventional medical doctors. According to Simpson, if used properly, high quality doses of hemp oil "…can provide a solution that will be of great help to mankind in alleviating this situation."

For several years Simpson has grown specific varieties of *Cannabis sativa* and treated patients suffering from the effects of chemotherapy free of charge. He states, "Through my experience I have found that there is nothing more effective or more harmless that can reduce the damage caused by radiation [chemotherapy]. I have seen patients who were suffering from cancer that were badly damaged by the effects of radiation treatments, were burned so badly by its effects that their skin looked like red leather. After ingesting the oil their skin went back to its normal healthy state and the radiation burns disappeared completely. If the oil can do this for someone who was badly damaged by so-called medical treatments, would its use not be effective to combat and cure the effects of radiation now emanating from Japan?"

Simpson contends there are many reasons why medicinal use of *Cannabis* should never have been restricted in the first place, but that chemotherapy and the associated diagnostic technologies represent a booming financial windfall, which continues to reap millions of dollars worldwide. Videos documenting Simpson's work in curing different forms of cancer and other conditions are available online as well as his new e-book (Simpson, 2009, 2012). His website can be found at: www.phoenixtears.ca

Major Breakthrough in Stem Cell Therapy

In early 2014 scientists heralded a "major scientific breakthrough" which had the potential to open up a new form of personalized medicine. Scientists from the Riken Centre for Developmental Biology in Japan discovered a new method for producing stem cells from red blood cells, simply by exposing the blood cells to acid. Stem cell research is important because stem cells have the ability to transform into any other type of cell in the body – including muscle, heart, nerve, liver and other organs. Accordingly, stem cell research has recently emerged as a new field in medicine, which has the potential to regenerate organs and tissues for the human body. Studies are already being conducted to evaluate the potential of using stem cells to heal heart, brain and eye tissue. These new studies also effectively eliminate any ethical issues regarding the use of embryos and umbilical cord tissue as sources for stem cells.

According to Haruko Obokata of the Riken Centre, "It's exciting to think about the new possibilities these findings offer us, not only in regenerative medicine, but in cancer [research] as well." The breakthrough was achieved using mouse blood cells but research is presently underway to achieve similar results with human blood cells. Researchers had developed methods to produce stem cells from skin cells, but in using this approach to heal age-related macular degeneration, for example, the time for culturing a patient's skin cells into stem cells suitable for injection into the eye takes ten months and is very expensive. The acid bath method can make stem cells available in a relatively short time and the treatment is much less expensive (Gallagher, 2014).

Possession Therapy

Possession Therapy (Exorcism), often referred to as "Spirit Release Therapy," is an area of Future-Science Medicine, which involves the removal of negative entities. I have added this section because over the years in dealing with clients, I have found that this is a much more common problem than most people realize and that many parents have experienced this problem of "unwanted entities" taking over a child's personality. The Catholic Church has openly acknowledged entity possession and exorcism has been routinely practiced by specially trained priests. Throughout the Church's history it has identified "demon-possessed individuals" and has thus developed specific rituals involving prayer and the use of holy water to get rid of them. Although it is not common knowledge until fairly recently newborn infants were exorcised as part of their christening ritual. The Catholic Church can thus provide counseling to Catholic families and those of other faiths who are interested (Ontario Consultants on Religious Tolerance, 2010).

Possession Therapy has recently emerged as a legitimate therapeutic practice, which focuses on identifying entity possession and treating patients who suffer from possession by discarnate or incarnate entities. The practice has proved especially effective in helping multiple personality disorders and is well-documented in the psychological literature. Since Entity Possession *can* be life-threatening it needs to be taken seriously.

For those wishing more information, a comprehensive overview of Spirit Releasement Therapy is presented by Dr. William Baldwin (Baldwin, 2009). Additional information and a list of practicing therapists can be found at The Association for Spirit Possession Research and Therapy website at: www.spiritpossesson.org Information on professional training and certification is available at: www.spiritreleasement.org/chr/trainings/basicsrt.html

Accepting Responsibility, Overcoming Tribal Beliefs and Healing Ourselves through Our Thoughts and Feelings

When working in the quantum field, *thinking* does not necessarily make it so. Effective healing and transformation also requires *walking the talk*. We need to fully commit to what we are saying and doing.

As children many of us were not unconditionally loved. This set us up for a lifestyle of "doing, doing, doing" in order to gain love and respect from others. It is important to understand that unconditional love can *never* be earned. We must learn to give it to ourselves. Crystal Nani in her book *Diary of a Medical Intuitive* provides the following insights on how to empower effective healing and transformation: "Your positive affirmations are indeed a good thing. I have clients who come in with illnesses who look in the mirror many times a day and affirm that they are getting healthy. Through their work with me, they have changed the energy behind their words. They believe them. This has resulted in chemotherapy patients raising their white blood cell counts, seeing their tumors shrink and having their blood tests return to normal. When they mean it, it works" (Nani, 2004).

Human/Extraterrestrial Healing

In her book, *Celestial Healing: Close Encounters That Heal*, author Virginia Aronson relates how in 1997 she began interviewing individuals who claimed to have experienced healing visits from extraterrestrial entities. "It was easier than I thought it would be to find such people, everyday people like ourselves, normal folk who have experienced some very abnormal things – and found themselves healed afterward. I was surprised that, mainly through word of mouth, I was able to meet over a dozen people within a short period of time. I also found that a number of these people had become healers themselves, an ability they attributed to their contact with the otherworldly beings." She relates how she met other individuals with unusual healing skills who claimed to be working with "extraterrestrial healing energies" and were using these skills to heal others (Aronson, 1999).

Israel's Extraterrestrial Healing Clinic

As improbable as it seems, a healing clinic was set up in Israel in 1994, where human psychic healers and teams of "ET doctors" developed a working relationship to diagnose and heal patients using advanced extraterrestrial technologies. The clinic was established by computer engineer and intuitive healer, Adrian Dvir. Dvir first became aware of his healing abilities when he attended an institute for spiritual healing and went on to complete degrees in Energetic Healing and Advanced Spiritual Healing.

According to Dvir, a group of extraterrestrial healers contacted him through well-known mystic and clairvoyant, Haya Levy. This ET group asked Dvir to work with them to create an etheric ET healing clinic in a space adjacent to his home in Israel. Dvir agreed, and the clinic was apparently built as promised. Levy assisted in the development of the clinic by functioning as a medium through which the ET doctors applied their advanced treatment protocols. Dvir claims he was contacted by an alien who identified himself as "X3" and told him that he had arrived to head up the medical team that worked with his friend Haya Levy.

Although patients who have received treatment at this clinic apparently rarely "see" the alien healers about 80 percent claim to have experienced strange tingling sensations during the procedures. The treatments, which are carried out in the higher realms, are apparently non-invasive and harmless. Since the clinic was established thousands of patients have been cured. Excellent results were reported to have been achieved with medical conditions including cardiovascular diseases, liver malfunction, immune system disorders, diabetes, orthopedic conditions and malignant cancer. During these medical procedures patients were not asked to believe in anything, but to simply give permission for the treatments to take place.

In 2003 Dvir published *Healing, Entities, and Aliens*, which provides extensive documentation of the many patient case histories at the clinic, as well as transcripts of conversations that took place between Dvir and the alien doctors. The medical procedures are summarized as follows by Journalist, Chani Solomon in the foreword for Dvir's book: "I was present at several of the treatment sessions presided over by the special team of aliens under the supervision of X3 and the other teams. I saw the tremendous improvement in the health of the subjects who were treated – results that are simply remarkable and fill one with wonder! The day can't be far off when our circle of knowledge expands to encompass awareness and understanding of these processes" (Dvir, 2003).

Adrian Dvir passed away in 2004 but apparently several ET/human clinics continue to operate in Israel and other parts of the world. Information on the Israel ET/human clinic can be found on Dvir's website at: www.adriandvir.com

In the U.S. an ET medical clinic was established by psychic healer Jackie Salvitti, who has a clinic in Las Vegas, NV and claims she is carrying on the work begun by Dvir (www.ethealing.com).

Final Thoughts on Future-Science Medicine

Future-Science Medicine offers exciting possibilities for creating synergistic teams to collaborate the abilities of psychic and intuitive healers with conventional medical technologies. This concept is perhaps best expressed as follows: "In this century, wellness and illness will be viewed as energy dynamics that are greatly under our control, and energy work will emerge as a chief means to achieve that desired state of personal balance. Energy work is exactly what the name implies: *harnessing energy to heal the body*. When you consider that everything in our universe, tangible or intangible, is composed of energy – your body, your thoughts, the chair you're sitting in, the sunlight streaming through your windows, the air you're breathing – there's little wonder this work can be transformational" (Christensen and Hillier, 2000).

MASTER KEY 16
Conscientious Biotechnology
Repelling Viruses, Reviving Mammoths, Cloning "Supertrees"

> "We already know how to clone animals.
> We are bioengineering new plants.
> Within a few years we will have sequenced the human genome
> (The DNA blueprint for a person).
> Soon, we will be able to grow replacement organs.
> By mid-century we may know how to 'engineer' new kinds of people.
> Such knowledge and power carries enormous responsibility,
> which the world has hardly begun to contemplate."
>
> *Professor M. G. Morgan, Head of Engineering and Public Policy,*
> *Carnegie Mellon University*

Through advanced applications of cutting-edge technology it has now become routine to re-program microbes to manufacture plastics, biofuels, vaccines and antibiotics. Microorganisms have also been genetically engineered to detect arsenic in drinking water, destroy cancer cells and store digital data in their DNA – creating a sort of "biological flash drive." At some point in the near future we may also be able to create microbial and human cells that are more resistant to viruses. It is also theoretically possible to bring extinct species like the wooly mammoth back to life. I have also included examples of "cloning" since it represents a natural step up from the natural selection process, but is not considered to be genetic engineering technology.

Certain applications of synthetic biology have already proven their scientific and commercial viability. For example plastic is usually made from petroleum-based hydrocarbons. Beginning in the early 1990's molecular biologist Oliver Peoples began searching for ways to alter a microorganism's metabolism so it could feed on biological feedstock and ferment it into the bioplastic compound polyhydroxybutyrate (PHB). Over the next 17 years, despite being ridiculed by several chemical companies he had approached, Peoples came up with a stain of proprietary microbe that could transform corn sugar into PHB plastics. This process is similar to the method used for brewing beer.

From an environmental perspective, even more significant, is a similar process, which involves using cyanobacteria (blue-green bacteria) to create biofuels that are virtually indistinguishable from petroleum-based fuels. The added bonus of this process is that these photosynthetic bacteria use sunlight, carbon dioxide and brackish water and convert them into alkenes – the basic molecular components of diesel oil. To this end Dr. George Church co-founded a company called Joule Unlimited, which has contracted with Audi to produce biological diesel fuel (biodiesel) at a price of U.S. 1.28 per gallon.

The concept of producing biodiesel fuel for this price is indeed a remarkable achievement. However it pales in comparison with the possibilities for using synthetic biology techniques to wipe out viral diseases. Viruses wreak their havoc by entering the cells of the host organism and then using the host's cellular mechanisms to replicate themselves – thus creating new virus particles that can kill the host organism. Viruses can accomplish this because they share the same genetic code as the host. Thus if we could slightly change the genetic code of the host cells, as well as that of the cellular machinery that reads and replicates the viral DNA the virus could no longer replicate.

Resurrecting extinct animals has already been accomplished to a limited degree. For example, in Spain in 2003 a falling tree killed a 13-year-old female goat named Celia. Celia was the last known living Pyrenean Ibex, a subspecies of wild mountain goat, so her species was essentially extinct. Fortunately, a Spanish biologist had taken skin samples from her ears and stored them in liquid nitrogen to preserve her genetic line. He and his associates removed the nucleus from one of the ear cells and transferred it into an egg cell of a domestic goat. They then implanted the fertilized egg into a surrogate mother goat that subsequently gave birth to a live Pyrenean Ibex.

Bringing back a wooly mammoth would be a much greater challenge since their remains are far older and the DNA would thus be more damaged. Thus, reviving a mammoth might require the reconstruction of its normal gene sequence (genome) by introducing gene sequences that researchers had already prepared.

Synthetic biology has its naysayers, but as with any science or technology it must be done with *integrity* and the possible environmental consequences must be taken into *careful* consideration. Synthetic biology is not a magical cure-all, but given that we can create plastic from plant material and diesel fuel from bacteria, the science offers possibilities for making us immune to viruses as well as for resurrecting extinct species. "It's a technology of unprecedented scope, power and promise. The challenge is to develop it safely and responsibly, with the public as a full partner" (Church and Regis, 2012).

Amazing Test Tube Forests

We are witnessing the dawning of a new era, where bio-engineered forestry and its associated technologies have the potential to revolutionize the $750-billion-a-year global forest industry – thus transforming forest landscapes the world over. Cellfor, Inc., a company in British Columbia, Canada worked for twelve years to develop a revolutionary new technology for creating fast-growing, genetically engineered "super-trees."

Cellfor's super-trees begin their existence as tiny greenish-brown dots pressed into nutrient cakes in glass Petri dishes. These tiny tree embryos are samples taken from some of the finest specimens of Douglas fir ever found in the wild. Douglas firs blanket much of the US Pacific Northwest. These magnificent trees tower to heights of up to 200 feet and are thus highly prized by lumber companies for their straight, knot-free wood.

According to Cellfor's president, Christopher Worthy, "What happens when you plant these trees is strikingly different from what happens when you plant ordinary unimproved trees." Cellfor's proprietary technologies also offered fascinating new possibilities for bypassing traditional seed orchards by creating millions of copies (clones) from a single prototypical seed and storing them cryogenically. Traditional tree farms have *not* previously used selective breeding to improve forest trees – a technique which has radically transformed most of our domesticated plants and animals for over thousands of years. Since cuttings do not easily propagate forest trees, forest biotechnology was considered an impractical and economically unfeasible option prior to Cellfor's breakthrough technology.

Cellfor developed a unique set of clonal technologies, which have the potential to increase wood production by as much as 60 percent from one generation to the next! Through improved breeding and management techniques foresters have been able to increase conventional lumber production by a factor of ten or more – all *without* genetic engineering. Cellfor's new techniques thus offer exciting possibilities for designing trees from the ground up. Researchers believe that in the next few years, companies like Cellfor will be able to create entirely new types of trees. These "supertrees" will be fast-growing, relatively knot-free and nearly branchless, making them ideally suited for lumber products. Such modified super-trees could be grown in intensively managed "super-forests." These supertrees could also be genetically engineered to produce alcohol or almost any other chemical directly by using solar energy and carbon dioxide (Mann, 2002).

As is often the case with start-up companies Cellfor was unable to raise the necessary capital to keep its company solvent and filed for Bankruptcy in 2012. Cellfor's breakthrough concepts for improving and restoring global forest resources will continue to be developed and implemented, since their proprietary technology and assets were subsequently acquired by the Canadian company, ArborGen. According to ArborGen's CEO, Andrew Baum, "The acquisition of Cellfor's seed embryos will allow ArborGen to accelerate and broaden its pine product offerings to commercial forestry customers." ArborGen has nurseries throughout the Southeastern U.S. and offices in Brazil and Australia. The company is said to be the world's largest supplier of seedlings and accounts for nearly 40 percent of the U.S. pine seedling market (Kearney, 2012).

Restoring Our Natural Ecological Heritage by Cloning Redwoods

In San Geronimo, California experienced tree-climbers have been taking small samples from the tops of some of Marin County's old-growth redwoods. The objective of this innovative project is to create new saplings from these "natural treasures" to help restore the legacy of America's former majestic forest resources. The concept of "cloning" is not new for plants and trees, as propagation from shoots and cuttings has been routine for centuries in orchards, vineyards and arboretums. The technique of cloning from sample bark slices or shoots *is* relatively new but had not previously been used for redwoods. In the redwood restoration project cloning is accomplished by dipping a cutting from a tree-top shoot into a rooting hormone solution then planting the shoot in a special moisture-controlled "fog chamber." As of the end of 2007 about 900 redwood cuttings had been rooted. According to the researchers involved it takes only about 20 of the seedlings to reforest one acre.

The project's director Dr. William J. Libby is a professor of forests and genetics at the University of California, Berkeley and a board member of the Save-the-Redwoods League. He has been helping to establish cloned redwood forests in England, France and New Zealand since the early 1980's. The California Redwood Restoration Project was organized by the non-profit Michigan-based Champion Tree Project International – an organization that has been instrumental in cloning many of the largest and oldest living trees in America. (www.nytimes.com/2007/11/27/science/27redw.html)

Producing Human Stem Cells Without Embryos

Two separate groups of scientists would seem to have achieved one of regenerative medicine's holy grails – the ability to reprogram mature human cells so they behave like embryonic stem cells and transform themselves into different types of adult cells for healing and regeneration. One important application of this process is in the area of tissue and organ transplants, as it helps eliminate rejection problems in transplant recipients. This new technique also removes the major ethical and religious objections initially associated with stem cell research since it does not involve the use of human fetus or umbilical tissues.

Over a decade has passed since the first lab-produced stem cells were created. At that time, stem cell research faced determined resistance from religious and fundamentalist groups. According to Doug Melton, Director of the Harvard Stem Cell Institute, "I'd welcome this other method because it's easier to obtain the material and doesn't raise ethical questions that some find troubling." He went on to say: "…using this other approach should enormously increase the amount of funding available for the research."

Two separate research groups, one led by Shinya Yamanaka at Kyoto University in Japan, the other led by James Thompson and Junying Yu at the University of Wisconsin produced similar end results. Both teams engineered stem cells to produce four different genes. Both groups also discovered that re-programmed pluripotent cells derived from individual stem cells could be used for tissue transplants without risking immune system rejection. Although the new stem cells appear to look and act like stem cells derived from embryonic stem cells it is not yet clear just how similar they are. Initial results from Yamanaka's lab have indicated that some differences do exist. Screening the expression of some 30,000 different genes suggested that the newly created pluripotent cells are similar but not identical to regular embryonic stem cells.

When implanted in mice, pluripotent cells generated masses of tissue containing multiple differentiated cells…a standard test for cell pluripotency. Using similar research protocols with embryonic stem cells Yamanaka's team demonstrated that the pluripotent cells could differentiate into muscle and nerve cells. Experimental results were published in the online journals, *Cell* and *Science*. James Thompson offered the following encouraging statement: "My personal barometer of optimism has gone up a lot." He added, "I think young investigators avoided getting into this field because of the ethical issues…Now, I believe more and more labs will move to this [new] method" (Singer, 2007).

∞

Genetically Modified Salmon Awaiting FDA Approval for Human Consumption

AquaBounty, a company based in Nova Scotia, Canada has produced and patented the first genetically engineered salmon. The "AquaAdvantage Salmon" contains a growth hormone gene from Chinook salmon as well as a "genetic switch" from an eel-like fish called the ocean pout. The new hybrid salmon can reportedly grow to harvesting size in *half* the time when compared with conventional farmed salmon.

AquaBounty produces the salmon eggs at its facility on Prince Edward Island. The eggs are then shipped to Panama where the fish are raised to maturity. In response to concerns that the genetically engineered salmon could escape, compete for food or breed with wild stocks the FDA has gone on record, stating that the chances are extremely remote since the salmon would be raised on land in tanks fitted with multiple escape barriers. Even if some fish managed to escape the natural waters in the area would be too hot and salty for them to survive. Neither would crossbreeding with wild fish be a problem since the fish would be sterilized and thus unable to reproduce.

In 2010, the FDA concluded that the AquaAdvantage salmon was as safe to eat as ordinary salmon. Final approval by the agency remains stalled pending an "environmental assessment study." The new genetically engineered species would effectively cut costs, time-to-harvest, electricity, labor and water use by half. The environmental impact would also be significantly reduced.

The proposed final approval of AquaAdvantage salmon has met with the predictable knee-jerk reactions from environmental groups and others. For example, according to Andrew Kimbrell, executive director of the Center for Food Safety – a Washington, DC advocacy group which opposes farm biotechnology, "The G.E. salmon has no socially redeeming value. It's bad for the consumer, bad for the salmon industry and bad for the environment.

FDA's decision is premature and misguided" (Perrone, 2012).

In December of 2012 the FDA announced it had concluded that AquaAdvantage salmon were "...as safe as food from conventional Atlantic salmon." Since AquaAdvantage has already spent *over ten years* seeking approval, this announcement was another significant step towards final resolution. AquaBounty's CEO was conservatively optimistic although he did not hesitate to comment that certain Congressmen had attempted to block the FDA from approving its patented salmon (Pollack, 2012).

Biotechnology scientists have recently begun shifting their strategies to avoid getting entangled in the lengthy FDA approval process. For example, researchers at the University of California, Davis, have created a herd of genetically engineered goats that produce protein-enriched milk. By shipping their experimental herd to Brazil they hope to avoid becoming entangled in U.S. governmental regulations. Other researchers at Canada's University of Guelph have produced a genetically engineered pig that emits significantly less environmental pollutants in its manure...a potentially major environmental breakthrough for factory farms, which are a major source of environmental pollution. Although there are *no* real reasons why the "Enviropig" should not be acceptable for human consumption the scientists have withdrawn their application for FDA approval based on the AquaBounty experience (Chernoff, 2010). Other projects include a genetically altered bird-flu-resistant chicken, which is being developed at Scotland's University of Edinburgh (Perrone, op. cit.).

From the author's perspective genetic engineering is, in itself, *neither good nor bad*. Genetic engineering uses cutting edge technology to speed up a process that has been naturally occurring with living organisms over hundreds and thousands of years of geological time. Like it or not, the truth of the matter is that advances in biotechnology are already underway and there is no realistic way of stopping it since foreign researchers will continue to move ahead regardless of what the U.S. thinks. Countries like China and

India are pouring millions of dollars into genetic research in efforts to create safer, more cost-effective and nutritious foods. The environmental nightmare scenarios created by big corporate GMO production should be regarded as lessons on how biotechnology *should not* be used. Thus companies like Monsanto, motivated by greed and the desire for money and power, have in a sense, done us a favor by creating the environmental nightmare scenario which presently exists – with farmers in the neighborhood of GMO corn being unjustly persecuted for "growing" Monsanto's patented corn because the GM seeds were blown by the wind on to their fields. In my personal opinion I believe it is through *conscientious* genetic engineering that new solutions will be found for repairing the shameful damage, which has already occurred. Our presently listless FDA regulatory agency should be *leading* this effort instead of *interfering* with these entrepreneurial scientists and their venture capital investors, as they represent a precious resource for America's future. The concept of conscientious biotechnology implies that with this type of research it is *critical* to thoroughly investigate the possible unintended consequences that genetically modified products might cause. This would create win-win-win situations for scientists, investors and bureaucrats alike.

With regard to public concerns about labelling genetically modified ingredients in food products, so many GMO ingredients are already incorporated into most of the processed food Americans consume, that it is impossible for food producers to comply. I have a simple suggestion: Label GMO food products that contain *no* GMO ingredients with a red-slashed circle with the letters "GMO" inside. For those who are frustrated with the present situation it is important to remember that we still have the power to "vote with our wallets." Companies like Monsanto *are* concerned about their public image and at some level understand they cannot continue to survive and thrive unless consumers are buying and consuming their products. They *will* come around, albeit slowly, and begin producing products that their customers want. I would thus advise concerned readers to do their homework, as only time will tell how the GMO scenario will eventually be resolved.

Future-Science Applications of Conscientious Genetic Engineering

Responsible applications of genetic engineering and its associated technologies can thus be used to create superior varieties of organisms which are fast-growing, disease resistant, require reduced fertilizers and pesticides, and are more immune to the environmental stress from pollution and global climate change fluctuations. Ultimately, it should be possible to merge the areas of psychic technology, consciousness research and genetic engineering so that we humans will ultimately be able to understand and manipulate the genetic blueprints of life itself.

MASTER KEY 17
Super-Ecology: A New Future Paradigm for Sustainable Living

> "The time has arrived in Earth's evolutionary history
> for people of all nations to make a quantum shift
> from regional and national perspectives to a new planetary consciousness.
> Technology has reached the point where we can choose
> to either enhance, or systematically destroy
> the delicate beauty and intricate balance of our global biosphere
> for the present and future generations of humankind."
>
> *Elliott Maynard – "Visions for the Future of Humanity and Planet Earth"*
> *(Unpublished manuscript)*

Super-Ecology: Restoring Planet Earth's Natural Resources

"Super-Ecology" is a new paradigm for restoring the global biosphere. It combines our latest scientific technology and knowledge with time-honored wisdom and indigenous "ecosense." The ultimate objective is to improve the natural environment in every way possible. Super-ecology thus incorporates the following set of actions: 1) First, the environmental region under consideration must be granted "legal preservation status." 2) Next, measures need to be put into place to protect designated areas from indiscriminate resource plundering so they can revert to their former natural states. 3) Once protected, the natural preserves thus created constitute "biological archives," which can provide reference models for mature ecosystems, which exist in similar ecological zones of our planet. Ecological resorts, donations and fair usage fees would insure the protection, maintenance and perpetuation of these nature preserves. The objective would be to make these preserves self-sustaining, while, at the same time, recognizing their importance as environmental legacies for the generations yet to come.

Natural ecosystems can often be enhanced by simply removing trash and invasive plants. For example, in smaller wilderness parks, removing extraneous deadwood and underbrush serves to enhance the forest as a whole – much the way European parks and royal forests have been managed for centuries. Another simple approach to environmental enhancement would be to add missing trace minerals and organic mulch – created by pulverizing excess brush and dead wood. Since many soils are nutrient-deficient in one aspect or another, balancing out soil composition encourages robust and healthy forest growth. Crowded trees can be thinned to allow each tree or shrub sufficient sunlight and growing space. Conversely, diseased, weak or ecologically incompatible trees, shrubs and invasive species can be eliminated and replaced with trees, bushes, medicinal herbs and flower plantings which are compatible with the particular ecosystem. This would increase the synergistic vitality for each ecosystem as dictated by its own unique requirements of soil type, rainfall, latitude and altitude.

Super-ecological environments can be funded by cooperative efforts between governmental, scientific and charitable organizations. They can also be cooperatively owned and managed by groups of private individuals interested in supporting environmental restoration. Super-ecological preserves can be created and maintained for outdoor recreation, environmental education, eco-tourism and environmental housing. Other forested areas, can be designated as timber farms for the sustainable harvesting of timber resources.

In the future, enhanced environmental forest parks can be enclosed in urban "eco-domes." Such domed enclosures would allow for the creation of "forest parks," which can be integrated into the design of futuristic "eco-cities." Such eco-cities would help reconnect humans with nature, as they would combine the best of natural and urban life. Trees and vegetation would provide shade to cut cooling costs, remove carbon dioxide and produce oxygen to clean pollutants from the atmosphere. Within these isolated climate-controlled biodomes the hardiest and most desirable species from around the world could be brought together along with a selection of herbs, bushes, grasses, flowers and wildlife. Light quality, day length, temperature cycles, carbon dioxide concentrations and humidity could be controlled to provide optimum conditions for growth and vitality. In time, these forest parks could also become major food production centers by using standard permaculture methods and enclosed greenhouse systems for the year-round production of agricultural crops, algae and aquaculture products. The brilliant inventor and futurist Jacque Fresco in conjunction with "The Venus Project" have created numerous examples of futuristic architecture being tastefully integrated into the natural world. Jacque's amazing videos and designs can be found at his website: www.thevenusproject.com

The concept of super-ecology can be extended to include the marine environments, where artificial reefs and marine preserves can be created in areas where natural reefs would not normally occur. For example, coral gardens can be "seeded" and "weeded" using technology to assist Nature in maintaining an optimal balance for the health and diversity of each unique super-ecosystem. Super-reefs would attract scuba divers and underwater photographers who would bring much-needed eco-tourist dollars to these remote areas. Such ecological preserves would serve as "nurseries" for producing juvenile fish and invertebrates, which can then spread out to adjacent areas to enrich local fishing grounds.

Restorative Aquaculture: Strategies for Marine Environmental Resource Management

One of the key aspects of the Super-Ecology paradigm is Restorative Aquaculture. Restorative aquaculture can be defined as the protection and sustainable management of marine ecosystems such as sea-grass shoals, mangrove communities, oyster beds and coral reefs. Coastal ecosystems the world over are threatened by dredging, commercial development, sewage runoff and agricultural pollution. The protection of these biological habitats is important, since they function as nursery grounds for countless species of juvenile fishes, crustaceans and bivalves. In addition to being incredibly rich biological habitats these coastal ecosystems also play vital roles in protecting exposed shorelines from erosion by dissipating wave energy from major storms.

Other ecological habitats that need protection include the so-called deep reefs, which are found at depths exceeding the limits of recreational scuba divers. These unique deep-sea gardens exist in cold water with little light penetration from the surface. The delicate corals found on these reefs differ from their tropical counterparts in that they lack the symbiotic algae (Zooxanthellae) that manufacture food for them. Because they are hidden from the public eye these recently discovered deep reefs have already suffered significant damage from years of unregulated commercial bottom trawling. Since deep reef organisms are delicate and grow very slowly it is important to understand they represent biological repositories, which are still relatively unknown to science. International agreements are thus needed to chart known deep reef areas and designate specific "no-trawl zones" so these delicate ecosystems can be preserved.

Sadly, over the past few decades the world's coral reefs have already lost most of their large fish species due to spear fishing and destructive fishing methods which use dynamite or poisons to collect specimens for the global aquarium trade. This tragedy has occurred because these "heritage species" have been routinely considered to be the most desirable and lucrative catches. This shift occurred so rapidly that evolution could not begin to keep pace with the new sports fishing and spearfishing technologies as they continued to develop. In my own personal experience as a graduate student in South Florida in the early 1960's I remember personally witnessing a spear fishing contest where large groupers, barracudas and other "trophy" reef fish were displayed like cordwood stacked on the shore at the end of the contest. During that era, I was privileged to experience close encounters with giant groupers over six feet long and weighing up to 700 pounds on the deep reefs of the Florida Keys – usually at depths of 100 feet or more. These giant fish, often called "Goliath Groupers" take many years to grow to their maximum size so catching them has been prohibited since 1990. In some of the marine preserves these "sequoias of the sea" have been protected from anglers and spear fishermen (in Cuba, for example) they have become tame enough to hand feed. Since these gentle giants have had no reason to be afraid of humans they provide unforgettable close-up encounter opportunities for underwater adventurers, children and photographers alike.

Specialized fisheries for Asian live fish restaurants and pharmaceutical markets also continue to take a steady toll on remote and pristine reefs the world over. The indiscriminate collection of exotic shells and marine specimens for local craft industries in tropical island areas of the world is another problem, which is rarely addressed. Over-exploitation of these shells has pushed some exotic conchs and cowries to the brink of extinction by reducing breeding populations and destroying their fragile ecosystems. Solving these problems begins with: 1) Establishing environmental awareness programs in elementary schools and local community organizations; 2) Insuring that indigenous people fully understand the importance of ecosensible approaches to preserving the natural environment around them so it can continue to produce food, jobs and ecotourism dollars for their local economies; and 3) Creating programs which encourage indigenous people and local businesses to move from being "part of the problem" to being "part of the solution."

Sustainable Aquafarms and "Environmental Tithing"

One effective solution for environmental restoration involves the use of intensive aquaculture to breed endangered or highly desirable marine organisms in captivity by growing out the young in protected aquaculture systems. Intensified artificial environments provide optimal conditions for the enhanced growth and survival of most aquatic species. The concept of "environmental tithing" is a simple requirement that ten percent of the juvenile organisms grown in small family aquafarms be released back into the natural environment to "seed" nearby areas with commercially valuable fish. Another approach to enriching the local fisheries involves using aquaculture techniques for growing fishes to adult size then releasing them back into the wild as breeding adults. In protected "no fishing zones" these adult specimens would have the opportunity to begin producing large numbers of offspring, thus re-establishing critical breeding populations in overfished areas. Larvae from these protected fishes would also disperse to repopulate nearby subsistence and commercial fishing areas.

Environmental enhancement programs can be set up to fund aquatic nurseries with support and assistance from government and environmental organizations. This type of program would provide indigenous people with sustainable cottage industries that would enhance (rather than deplete) local fisheries and restore regional sustainability. Small-scale "aquafarms" could focus on producing the most desirable tropical fishes, exotic shrimps, anemones, mollusks and corals for the global aquarium trade. This way, environmentally sustainable business enterprises could be established for local families and communities.

Selected marine areas in every region of the world should thus be set-aside as marine parks, environmental education centers and biological banks. These national marine parks would "seed" nearby areas for commercial and recreational fishing. Other protected areas should be created for scuba divers, snorkelers and underwater photographers. The key to success with restorative aquaculture lies in the fact that marine organisms produce large numbers of eggs to compensate for predation and other normal losses in the wild. Since restorative aquaculture can effectively boost natural survival rates by factors of 100 to 1,000 times or more this concept represents a "magical tool" for restoring areas where environmental damage and overfishing have decimated natural populations of endangered marine species.

Great Barrier Reef Corals are Disappearing at an Alarming Rate

Following extensive studies of 214 coral reefs scientists discovered some alarming news. Coral cover of the reefs surveyed had been reduced by 50 percent since surveys in 1998. Researchers also found that coral cover had dropped by over 50 percent. Coral reef expert John Bruno sees these findings as "really grim." He stated, "In 2007 we first sounded the alarm that the Great Barrier Reef and Pacific reefs in general were not as pristine and resilient as a lot of people wanted to believe."

Researchers attributed the massive coral declines to severe tropical storms, crown-of-thorns starfish predation and coral bleaching, which accounted for 48, 42 and 10 percent of the declines respectively. The scientists suspect that the crown-of-thorns starfish is an indicator of poor water quality, based on the fact that the remote reefs studied had minimal starfish predation and *no* overall declines. Since nutrient-rich water encourages plankton blooms which starfish larvae thrive on the scientists theorized that if fertilizer runoff and other nutrient-rich pollution sources can be reduced, then starfish populations would decline and coral growth would recover at a rate of about one percent per year. Bruno expressed his feelings as follows: "The impact of starfish on the reef is striking. They are huge and scary beasts. They move in massive waves down the Great Barrier Reef like a plague" (Llanos, 2012).

Super-Ecology Strategies for Coral Reef Restoration

A company called Applied Marine Technologies on the Caribbean Island of Dominica has developed a technology for restoring endangered coral reefs. AMT established the first commercial "coral farm" in the world. Founder-Director Alan Lowe has been coral farming for two decades. His company consists of a dedicated team of engineers, divers and coral specialists. Their objective is to cultivate coral under controlled aquaculture conditions and replant small pieces in reef areas that have been environmentally stressed, damaged or destroyed.

Since corals reproduce asexually, each small piece of a coral tree transplanted to a damaged area can grow into a mature colony. Corals also reproduce sexually once or twice a year. They produce up to half a million eggs of which only a few survive in the natural environment. Under controlled aquaculture conditions survival percentages can be increased exponentially and used to re-seed damaged reef areas. AMT's research suggests that a few hundred pieces of cultured coral, when transplanted to a damaged reef structure, can give the reef a 165-year head start compared with a reef that goes through a natural recovery process (Whitford, 2000, Onion, 2013).

Another innovative approach to coral reef restoration uses a type of 3D Printing Technology. Overfishing, pollution and climate change have created environmental pressures on coral reefs worldwide. Depending on the region, average coral decline ranges from 40 to 80 percent. Recent studies predict even more drastic declines, mainly due to ocean acidification, which makes it more difficult for corals to grow. We have reached the point where some scientists feel that coral reefs could completely disappear in the lifetime of a child born today.

In a concerted effort to stimulate coral growth and provide suitable structures for marine organisms another group called "Reef Arabia" has been using concrete molds to create artificial reef structures in the Persian Gulf. In this context it is important to understand that artificial reef structures are not only good for corals, but provide a substrate for larval invertebrates, as well as food and shelter from predation for juvenile fishes and a host of other beneficial marine organisms. Reef Arabia has already installed three thousand "reef balls" in depleted areas of the Gulf. Reef balls are concrete half-domes with holes similar to Swiss cheese. An inflatable bladder allows the concrete structures to be towed to a desired location. The air is then let out of the bladder and the reef ball sinks to the ocean floor. The deflated air bladder is then pulled out through one of the openings in the concrete dome and can be used over and over again. Thousands of reef balls have been built and distributed by environmental groups worldwide. Even "memorial reefs" can be created by incorporating the ashes of a loved one into the concrete reef balls. (www.reefball.org)

Reef Arabia has taken the reef ball concept one step farther. Partnering with a UK company called D-Shape the group has begun using 3D printing technology to create more refined reef formations. According to David Lennon, member of Reef Arabia and Director of Sustainable Oceans International, "With 3D printing we can get closer to natural design because of its ability to produce very organic shapes and lay down material similar to how nature does it." Unlike reef balls, the 3D printed reefs are composed of a patented sandstone material that resembles the composition of living coral. The rough texture attracts marine species that use the structure as a refuge and horizontal surfaces are ideal for attracting coral larvae. The expectation is that this kind of substrate will help increase biodiversity, a key for making these ecosystems more resistant to climate change. The printed prototypes take about a week to fabricate although the actual printing process takes only about a day. When production is ramped up, Reef Arabia expects to print four reef units at a time.

According to Michael Webster, Executive Director of the Coral Reef Alliance, most people do not realize that corals are the organisms that build the calcareous framework for most coral reefs. He feels that what *really* needs to be addressed are the conditions which are adversely impacting coral reef communities worldwide – mainly climate change, pollution, overfishing and ocean acidification. Once these negative factors can be reduced or eliminated, artificial structures such as reef balls and printed reefs will be able to establish themselves so that corals and other marine organisms can multiply and flourish (Antoniades, 2013).

Seeding Ocean Currents with Powdered Iron to Create Plankton Blooms and Enrich Food Chains for Ocean Fisheries

Several years ago I attended a new energy conference hosted by the Integrity Research Institute in Washington, DC. One of the sessions was presented by marine scientist, Russ George, who outlined his plans for seeding selected areas of the ocean with powdered iron – in areas where iron was the limiting factor for phytoplankton blooms. He hypothesized that other organisms in the food chain would consume this phytoplankton and ultimately sink to the deep ocean floor, thus effectively removing carbon from the atmosphere. Russ came up with the idea of using this system to assist corporations who were at that time pressured to reduce their carbon emissions. Planktos, Inc. was thus established as a way to expand this new frontier and create a win-win situation for sequestering carbon on the ocean floor.

To test their hypothesis Russ and his crew set out in their research ship *Weatherbird* and proceeded to a section of the South Pacific Ocean where they ran a series of large-scale experiments. This research was based on the results of 11 different powdered iron-seeding experiments, which had been accomplished earlier and was validated through NASA satellite imagery. Planktos CEO William Coleman maintained that phytoplankton productivity had declined by a factor of as much as 14 percent over the past 20 years due to major shifts in ocean weather patterns. He also suggested that recent changes in soil management practices had further reduced the amounts of iron dust required for healthy plankton blooms. This shift, in addition to the elevated levels of carbon dioxide in the atmosphere from the burning of fossil fuels, has turned the oceans into "a slightly acidic carbonated beverage."

The focus of the Planktos research program was to demonstrate that significant amounts of atmospheric carbon would be consumed by these experimentally induced plankton blooms and thus become biologically locked-up (sequestered) by falling to the deep ocean floors, rather than being cycled back to the atmosphere. Planktos estimated that about 25 percent of the atmospheric carbon is sequestered when phytoplankton dies and falls below depths of 1,600 feet. Coleman also hypothesized that restoring algae levels to where they were 30 years ago could sequester about 50 percent of the carbon produced from burning of fossil fuels and other sources worldwide. Despite their confidence that these initial iron-seeding experiments would validate the concept of ocean iron-fertilization, Coleman admitted that the iron seeding might not work as intended. "If it doesn't, well, then we had the chance to demonstrate it and we all go home with better information and knowledge about how the oceans respond to iron restoration."

In common with previous iron-fertilization attempts Planktos had its share of skeptics and would-be spoilers. For example, Nicholas Meskhidze, biological oceanographer from the School of Marine, Earth and Atmospheric Sciences at North Carolina State University stated, "You can make that calculation on paper, but whether it actually works that way in the ocean I don't think is known." Another skeptic, biological oceanographer John Marra of Columbia University's Lamont-Doherty Earth Observatory summed up the reasoning behind this kind of skepticism: "It seems like every time [researchers] have done one of these iron-enriching experiments, they get different results" (Sergo, 2007).

An update of Russ George's research progress appeared in a *New Scientist* article published in 2012. By that time their progress had been impeded by unexpected interference by Greenpeace. Unfortunately the confrontation with Greenpeace occurred at a critical time when the *Weatherbird* was scheduled to take on a supply of powered iron for their next experiment. At about this same time, a major downturn in the stock markets also effectively shut off the ongoing funding sources. Lack of funds to operate the *Weatherbird* thus forced the Planktos program to terminate its operations. If this were not enough, all respondents to a questionnaire in *New Scientist* agreed that, despite George's good intentions to create a carbon trading program any further iron-seeding experiments should be run for the public benefit, and *not* on a profit basis. In addition to these unanticipated hurdles, in 2008 an international treaty called the London Protocol was drafted ruling that the practice of accruing and selling carbon credits should not be allowed (Marshall, 2012). As a result of these unfortunate events, Planktos filed for bankruptcy in 2008.

In 2013, in an attempt to sidestep international resistance to powdered iron fertilization, a team of marine scientists from the UK, Norway and South Africa, studied the effects of Iceland's Eyjafjallajökull Volcano. Results from their studies determined that particulate iron from the volcanic ash clouds *did* indeed trigger dense plankton blooms downwind from the volcano. Due to a lack of nitrogen, however, this lush bloom died out quickly. Instead of treating this research as a solid piece of the ocean fertilization puzzle, another knee-jerk media backlash came from The Climate News Network who were quick to label these results "a blow for geoengineering supporters" (Diep, 2013).

In June of 2013 a new wrinkle was revealed in the science of ocean fertilization. A research paper was published which highlighted the role of *diatoms* as agents for carbon sequestration. Diatoms are single-celled photosynthetic plankton organisms that have a skeleton composed of silica. This silica skeleton makes them difficult to ingest for larval fishes and crustaceans. The researchers studied samples taken along the coast of Western Antarctica. Using x-ray diffraction technology they determined that diatoms *were* indeed iron transporters. This added yet another key factor to the ocean fertilization equation.

In 2012 Russ George secured a contract with the Haida Tribe in British Columbia to assist these indigenous people in bringing back the salmon runs they have relied on in the past. An agreement was drawn up with a new company formed for this purpose – The Haida Salmon Restoration Corporation (HSRC). The company based their project objectives on the fact that in the summer of 2008 Mt. Kasatoshi in Alaska's Aleutian Islands erupted, sending clouds of iron-containing volcanic ash over vast areas of the ocean. Following this event plankton blooms were generated across those areas downwind from the erupting volcano. In 2010, after years of progressively declining salmon runs, fishermen were rewarded with a record salmon run in British Columbia's Frasier River. According to Russ, "There are three volcanic events in the last 100 years, and we had record sockeye runs following those three volcanic dust events. That's pretty good data." This project would have seemed to have "the right stuff" for a serious research effort. HSRC researchers used satellite imaging, two observation gliders on loan from the Canadian Institute for Ocean Science and 20 autonomous drifter robots from NOAA (the U.S. National Oceanographic and Atmospheric Administration). Funding was provided by a $2.5 million grant from the

Haida Tribe (Biello, 2012a).

When the story of this venture went public the Haida Restoration Project quickly became a target for unreasonable criticism by governmental and supposedly "green" organizations that discovered they could create sensationalism by using creative journalistic spin tactics to feather their own nests. Even worse was the fact that organizations like Greenpeace, who had supported demonstrations against George's earlier ocean iron seeding efforts, were quick to jump on the bandwagon – when they should have been doing their scientific homework and supporting the project. Sensationalized articles that highlight this kind of "green mischief" include: "A Rogue Climate Experiment Outrages Scientists" (Fountain, 2012) and "Does Russ George Deserve a Nobel Prize or A Prison Sentence?" (Abrams, 2013). Examples of articles written in a more positive vein include: "Geoengineering Trial Hailed a Success" (Marshall, 2012).

Since I had studied marine chemistry and phytoplankton ecology at the Nova University Oceanographic Institute in Ft. Lauderdale, Florida I was favorably impressed with Russ George as well as his initial ocean iron-fertilization projects. Recently, I tracked down Russ, who remembered our initial conversations in Washington, DC. This is what he had to say concerning my questions with regard to his current status, the new element of diatom carbon sequestering, the possible role of tiny nanoplankton and his plans for the future: "As for the diatoms, they have always been on everyone in the field's list as great carbon sinkers. Your 'nanoplankton' are a challenge – undoubtedly important but difficult to study. Somewhere in the tens of thousands of collections I have there are tens of thousands of discoveries to be made in that regard, but as I am not the 'wealthy' businessman-villain as I have been portrayed, as my financial resources are limited." Russ added, "Of even greater interest to me of late is the report on the historic underestimate of the tiny fish in the mid-depth ocean that take part in nightly feeding migrations. The estimates of this biomass have just been revised upward to *ten times* the former estimate. These mid-depth ocean fish, along with their zooplankton cohorts, are responsible for transporting truly massive amounts of carbon into the deep ocean on a nightly basis."

Despite the irrational overreactions by the press and being unconscionably smeared by media pundits who should know better, Russ has somehow managed to keep his sense of humor in spite of being a major player in this modern scientific inquisition drama. "Like all pioneers I find I count far more arrows in my back than in my front. Naturally people around the world have taken note of the spectacular success of my 2012 project in bringing the largest catches of salmon in all of history to the NE Pacific." He went on to add: "It just works! The ocean can be replenished and restored and the fish come back. Naturally, one must conduct repeat experiments to solidify and refine the methods and technologies. That is a big job which I am working on diligently" (George, 2014). Updated information on Russ' recent activities in ocean nutrient enrichment is available at his website: www.russgeorge.net

Final Thoughts on Super-Ecology

By applying the principles of Super-Ecology global forest preserves and urban forest parks can be created using genetically enhanced trees, selected for their rapid growth, disease resistance, air-cleaning abilities and natural beauty. Marine hatchery and nursery farms can be established to seed the world's oceans with fast-growing fish and shellfish and to grow replacement corals for restoring reefs which have been damaged or destroyed by pollution and coral bleaching.

Programs can also be implemented to restore, monitor and manage global ocean fisheries, using satellite-imaging technology to monitor the "health" of marine ecosystems. Cooperative efforts between government and industry can help manage wild game populations and restore endangered species. In essence, super-ecology is simply a new approach for restoring damaged ecosystems, which applies cutting edge technologies to create and sustain a healthy balance between humans and the planetary biosphere.

MASTER KEY 18
Harmonic Attunement Technology: A New Method to Upgrade Human and Planetary Consciousness

> "We know from Gautama the Buddha,
> that there is a mysterious phenomenon which he termed the 'Buddhafield.'
> Two thousand five hundred years ago,
> there was an extraordinary following of over ten thousand monks
> gathered around him in the state of Bihar, India.
> He maintained that the intense energy field that was generated around him
> could start a chain reaction.
> It certainly seemed to, for, from all accounts,
> more disciples entered the natural enlightened state with him
> than with any other master at any time."
>
> *Yatri, 1988 – Unknown Man: The Mysterious Birth of a New Species*

Harmonic Attunement Technology: A New Paradigm that Integrates "Resonance" and "Consciousness Entrainment"

Harmonic Attunement Technology was created for the purpose of uplifting the human consciousness. This can be accomplished in our homes and workplaces, as well as in theaters, concert halls and stadiums. By combining consciousness technology with subtle-energy electronics it is possible to "tune" a room, building or stadium with intensified natural energies – thus creating an atmosphere of peace, harmony and happiness. In addition to *directing, enhancing* and *intensifying* these positive energies electronic devices can be designed to screen out the negative influence of manmade "electronic smog" produced by high voltage power lines, smart meters, wi-fi routers, cell phones, cell phone towers and computers.

The consciousness enhancing frequencies produced by harmonic attunement devices expand our normal physical, mental and intuitive abilities. Through the principle of sympathetic resonance, human energy fields tend to synchronize with the energy patterns from these devices. This augmentation effect offers fascinating possibilities for teams of individuals to function together symbiotically as a kind of *super organism*. This type of heightened consciousness promotes clarity of thought and allows us to more easily access and work with the quantum field. Within these electronically created states of consciousness breakthrough thinking can occur on a regular basis, since creative solutions to problems become easier to access and manifest in the physical realm.

Harmonic attunement technology is an area in the greater field of consciousness enhancement technology, which employs electronic devices to boost human brainpower, raise consciousness, heal and facilitate the natural intuitive process. Examples include light and sound machines, Ganzfield units, binaural sound recordings, biofeedback devices and bio-acoustic computer programs. These devices fall into the general category of "mind machines."

Harmonic attunement technology thus involves the creation of user-friendly devices, which represent electronic interfaces between consciousness and the higher realms. These "electronic gurus" provide instant feedback and training regimes for achieving higher states of consciousness – a process formerly requiring years of ascetic practice in traditional spiritual paths. Scientific studies have validated the effectiveness of mind machines for alleviating stress and anxiety, as witnessed by their use in the clinical treatment of substance abuse, depression and anxiety. Michael Hutchison's *Megabrain* book series provides an excellent overview of mind machines and their applications. He states, "We've convinced businesses to use these devices for stress reduction, schools for better learning curves and doctors for drug rehabilitation" (Hutchinson, 1996).

Back in the 1940's scientists discovered that human brain waves tended to synchronize with flashing light and sound frequencies. This led to the development of a series of sound- and light-entrainment mind machines. Biofeedback is an important aspect of any consciousness-enhancement technology, since it provides measurable feedback in the form of brainwave patterns that are displayed on a monitor. Such devices use electrodes placed on the scalp and other parts of the body to monitor these patterns. Through visual and auditory feedback they provide a convenient means to more easily achieve relaxation, healing and deeper meditative states. These devices can also help balance out the neural pathways between the two cerebral hemispheres of the brain. This results in improved brain function and a more stable and holistic awareness.

Dr. Sung W. Lee, MD and his associates at the BrainWell Center in Sedona, Arizona have developed an excellent biofeedback brain-training program. I have worked with Dr. Lee in the past to experience first-hand the positive results from this intensive brain-training program. In Dr. Lee's system, electrodes are fastened to various points on the scalp of the client. During a series of one-hour sessions, the client is led through a series of guided meditations with eyes closed. Other exercises involve eyes-open realtime visual feedback, where the client attempts to smooth out the different neural patterns, which are displayed on a flat-screen monitor. Nothing is forced, as the results of each session are thoroughly reviewed and any questions concerning the process are answered. These brain-training sessions are usually spaced out over a period of two weeks. [Since your consciousness is shifted, it is difficult to define specific benefits from the program. However, following my own two-week brain-training session I was able to add some 20,000 words to the manuscript for this book in a space of about ten days]. Information on Dr. Lee's programs and the BrainWell Center are available at: www.brainwellcenter.com

Although harmonic attunement technology is still in its early developmental stages it represents a powerful tool for transformation of the human consciousness. Other consciousness-enhancement devices are designed to produce electronic frequencies, which mimic the natural pulse of the Earth (Schumann Resonance). As mentioned earlier evidence suggests that this basic "earth-pulse" has a profound influence on the evolution and health of all life on our planet. As an interesting side note, physicist/engineer Bob Beck discovered that this was the same frequency, which was found to be dominant in the brain waves of psychics and mystics when they entered higher states of consciousness (Hunt, 1989).

It is important to point out that human/electronic field-enhancement devices are intended to create a field which is uplifting, healing and can serve as a "consciousness catalyst" for groups of people to work together more coherently and synergistically. In this same context, Dr. John Hagelin, Director of the Institute of Science and Technology at the Maharishi Institute of Management, has proposed the use of coherent meditation groups for creating "peace shields" and generally uplifting the human consciousness (Hagelin, 2002).

Through specific applications of electronics technology, it is thus possible to produce consciousness enhancement fields for concert venues, performers and audiences that create an atmosphere of peace, happiness and well-being. This technology can be expanded to include tuning the production technologies, which go into each event – lighting and sound systems and their associated electronic components.

The common denominator for consciousness enhancement devices is that they function partly in the quantum field – outside the boundaries of traditional scientific thinking. In 2003 I collaborated with a Russian scientist, who provided information on Russian psychic technologies, which had only recently become available to the western world. In his opinion Russian science was at that time about 30 years ahead of the U.S. with regard to psychic research and technologies for measuring and amplifying human psychic abilities. [This information had been kept secret, since the research was focused on military technology prior to that time]. After the fall of the iron curtain, some of this research was translated into English with the intent that its applications would be used for purposes of healing, consciousness enhancement and peace (Ivanenko, 2003).

Final Thoughts on Harmonic Attunement Technology

In accord with the principles of sympathetic resonance our energy fields tend to align themselves with the energy patterns of electronically generated fields. This entrainment effect offers the potential for groups of individuals to function together more synergistically, to uplift the general public consciousness and to promote international peace. By working in a quantum modality new insights and breakthroughs tend to surface more easily. In this kind of heightened consciousness state breakthroughs are not just *likely* to occur, but will *inevitably* occur as part of this new operational matrix.

Harmonic Attunement Technology devices and related mind machines thus offer a cost-effective means for relieving stress and anxiety in our fast-paced lifestyles. They can also help shield our brains and bodies from electronic smog. The use of these devices can be extended in homes, offices, schools, medical facilities, museums, concert halls and government buildings, with the objective of increasing energy levels, clarity of thinking and the general health and well-being for everyone concerned. Just imagine the possibilities!

MASTER KEY 19
Superfoods and Supernutrition

> "Eating is an essential part of all of our lives – we simply can't escape it.
> How we choose to experience food is up to us as individuals.
> We can think of food as pure physical nourishment –
> to give us the calories we need to function – to move, speak, and think.
> Or, we can expand our vision to encompass eating
> as something that connects us to our inner and outer landscapes –
> our emotions, mind, environment, people, animals, plants,
> earth, water, planet, and universe."
>
> *Deanna Minich, Ph.D., 2010 - www.foodandspirit.com*

The concept of Future-Science Nutrition begins where conventional nutritional science leaves off. Rather than measuring food in terms of caloric value or molecular composition future-science nutrition is based on the vibrational energy of foods – especially with regard to the positive effects of these foods on the subtle-energy fields associated with our human bodies.

Superfoods: What are They and Where Do They Come From?

Many of the natural high-energy foods most suitable for nurturing the physical and finer bodies come from the marine environment. Here, solar energy is converted via photosynthesis into highly nutritious foods, which are then passed along up the food chain. Examples include *finfish* such as trout, salmon, catfish and tilapia; *shellfish* such as lobsters, crabs, shrimps, clams, mussels and conchs; *seaweeds* like Irish moss, kelp and nori; and *algae* such as *Spirulina* and *Chlorella*.

Two examples of high-energy seafoods yet to be commercially developed include: *Marine yeasts* which can be grown in brackish water – or even crude oil; and *Krill,* two-inch long planktonic crustaceans which are the basic food for whales, penguins and fishes in Arctic and Antarctic waters. These hi-protein products are ideal for producing high-energy nutritional powders, which are a perfect food for humans, pets, food animals and aquaculture species alike.

Many types of seafood can also be raised to higher levels of nutritional excellence by growing them in controlled environmental aquaculture systems. Such "protein factories" would have significant advantages over conventional farms. For example, they can be located near consumer centers. This reduces transportation costs, produces local jobs and makes fresh food products available in urban areas.

By applying the principles of Future-Science Technology to environmentally controlled culture systems the basic nutritional value of the food products can be boosted to high levels. In these systems environmental factors such as temperature, pH, light spectrums, periodicity and intensity can all be optimized to produce the ultimate in designer foods.

When the crops are harvested, special handling and preservation techniques such as flash freezing and spray-drying can be employed to preserve the high-energy levels of the food products. Supernutritional supplements in tablet, powder or liquid form can be further potentized by the subtle-energy techniques described previously in this book.

The "Living Lunchbox" Concept

The Living Lunchbox Concept is an idea, which came to me as a Research Scientist at the Kuwait Institute for Scientific Research in 1980-81. This concept represents a new approach for creating ultra-high-energy superfoods, and focuses on the intensive culture of small food organisms to feed larval aquaculture fish and shrimp. The concept involves raising live food organisms with *optimal* nutritional diets so the farmed fish or shrimp that feed on them will have correspondingly higher growth and survival rates, and superior nutritional qualities that surpass their wild counterparts. Examples of live food species include the single-celled alga *Chlorella*, marine yeast (*Candida*), rotifers and brine shrimp (*Artemia*). Fish and shrimp produced in this way can also be considered to be "designer superfoods of the future."

Intensive aquaculture systems thus provide ideal environments for efficiently delivering high-powered nutrients to biological food organisms such as marine yeasts, single-celled algae, rotifers, and brine shrimp. These "living lunchboxes" provide live food of superior quality for larval and juvenile stages of shrimps, fishes and shellfish produced in aquaculture systems.

From a perspective of energetics it is most efficient to use natural solar-energy-driven photosynthesis to grow single-celled algae like *Spirulina*, which, even in its natural state, is a superfood composed of 70 percent protein and nine essential amino acids. Raising herbivorous fishes that feed on algae and phytoplankton is more energetically and ecologically efficient than raising carnivorous fish like trout, catfish or salmon, since these latter species are higher up on the biological food chain, and require commercially processed pelleted foods *made from ocean fish stocks* – thus energetically "robbing Peter, to pay Paul" (Maynard, 2009).

The fact that the growing cycle of these smaller food organisms is usually a matter of days or weeks gives algae-fed fish farmers a significant advantage over farmers raising carnivorous aquaculture species. This advantage is further magnified when we compare aquaculture grow-out cycle times to the land-based crop cycles corn or apples, or to the farm animal production cycles for poultry, cattle or hogs.

Designer Eggs: An Extension of the Living Lunchbox Concept

"Designer eggs" are standard eggs that have had their nutritional value boosted to meet health-conscious consumer demands. Designer eggs that contain higher concentrations of specific vitamins (usually vitamin A and E) are now available in most supermarkets. Since the diet of the laying hen influences the nutritional content of the eggs, hens fed with special diets of kelp, flax seed and canola oil produce eggs with lower saturated fat content and higher omega-3 fatty acid, Iodine and Vitamin E content. Designer eggs are becoming an increasingly popular choice in the US market for health-conscious consumers, as in the US alone designer eggs comprise nearly five percent of the five billion dollar egg market. On average, designer eggs cost about a dollar more than a dozen ordinary eggs. Designer eggs provide a good example of upgrading an established product to produce a superfood. (www.foodproductiondaily.com/Packaging/Designer-eggs-latest-US-trend)

"Cage-free eggs" are produced from hens that are allowed to run free in larger pens, eating natural foods such as bugs, seeds, berries and natural organic material in addition to their regular diet. Many of these chickens are also steroid and antibiotic-free, so the concept of cage-free eggs might best be compared to humans eating natural foods from their own gardens.

Although it is not commonly known, chickens are "natural mulching machines" and will eat a wide variety of table scraps in addition to insects and organic plant material. This makes them ideal in tropical areas of the world where they can be allowed to free-range on the tropical forest floor, along with small numbers of pigs, cows, ducks, geese and goats. Using sustainable management techniques these animals function to recycle all edible material while cycling their waste products back to the environment as organic fertilizer. Eggs, meat and dairy products from these free-range animals thus provide some of the most healthy and nutritious food products on the planet.

Producing Higher Quality Fish with Designer Diets

Growing farm-raised fish with special "designer diets" provides a new way to get highly nutritional foods to consumers who value high-energy foods and healthy lifestyles. According to a Purdue University research team, farmed fish fed with high fatty acid diets can provide significant nutritional benefits for those who consume them.

The Purdue scientists are currently focusing their research on a new type of fatty acid called conjugated linoleic acid (CLA), which medical researchers have found to be a key factor for prevention of diabetes and cancer. This is in line with recommendations by the National Academy of Sciences Institute of Medicine. According to Purdue researcher, Paul Brown, "We found that by adding CLA to fishes' diets we can get more of these fatty acids into fishes' tissues than is found in any other animal. Fish have always been the original and standard measure for good sources of omega-3, but now we find that we can introduce other fatty acids into the fish. Next, we must determine if there is an optimum ratio of omega-3 to omega-6 fatty acids that is healthy." Specialized diets for farm-raised fish can thus be formulated with nutritional additives like CLA and selected grains to produce "designer fish" that contain optimum concentrations of beneficial fatty acids and other nutrients. To this end, the Purdue scientists are now studying different fish species to determine just how much of these nutrients they retain when they are provided with special high-nutrient diets.

The ability to create "designer fish" which are superior in both taste and nutritional quality should provide significant advantages for the aquaculture industry in general, since maximum sustainable yields for global fisheries stocks were reached at the end of the 1980's. In view of the shrinking of world fisheries resources there is an urgent need to increase aquaculture production globally. In Brown's words, "We have to develop new aquaculture production that rivals global production of soybeans, pigs and chickens if we want to keep eating fish and shellfish."
(www.flmnh.ufl.edu/fish/innews/fishdiet2003.htm)

∞

Biodynamic Agriculture: A Natural Way to Produce Superfoods

Dr. Rudolf Steiner spent his career researching the forces that regulate life and plant growth. In 1924 he created the basic principles for biodynamic farming. Essentially, biodynamic farming is a unified approach to agriculture, which relates soil ecology to the forces and cycles of the natural world. Biodynamic farming regards maintenance of the soil, including its nutrients and living organisms, as a "living entity" which can be nurtured and sustained for centuries through the principles of organic farming. Maintaining a healthy and balanced soil helps prevent soil erosion and eliminates the need for chemical fertilizers and pesticides. This results in improved crop yields and high-energy foods for both humans and animals.

Biodynamic agriculture works to improve soils through proper humus management. This is accomplished through applications of organic manure and compost, which are in their optimum stages of fermentation. Other basic aspects include proper crop rotation and the use of cover crops and "green manure," which is produced by solar radiation from organic mulch. The resulting nutrient teas can be sprayed over the crops as liquid fertilizer. Other key aspects of biodynamic agriculture include planting two or more synergistically compatible crops together to avoid insect pests and thus provide more ecologically balanced gardens.

Biodynamic techniques also use compost to culture earthworms as well as certain herbs and beneficial microorganisms in the roots of legumes to aerate and enrich the soil. Composting techniques involve the layering of organic matter and the addition of beneficial microorganisms to accelerate the process. The end result is an organic fertilizer, which is ecologically balanced and nutritionally superior. (www.biodynamics.com/biodynamics.html)

In conclusion, biodynamic agriculture encompasses a new way of living, working and relating to Nature. The science is based on a combination of common-sense and a deep understanding of time-tested ecological principles. It also represents a more spiritual approach to the basic principles and methods of agriculture. "Biodynamic farmers have practiced the principles and methods of a sound, sustainable agriculture for over sixty years. They farm in many production zones and in many countries. Their work is grounded in an evolving understanding of the forces and substances that fashion living nature. Biodynamics offers workable answers to many pressing issues in contemporary agriculture" (Koepf, 1989).

Paramagnetic Agriculture as an Aspect of Superfood Production

The concept of paramagnetic agriculture was pioneered by Dr. Phil Callahan, Professor of Entomology at the University of Florida. According to Dr. Callahan, "Paramagnetism is the ability of a substance to collect or resonate to the magnetic fields of the Cosmos." Dr. Callahan's work is based on over 50 years of research on earth energies and their influences on health and agriculture. From his extensive work he concluded that it was essential to understand the energetic background of life itself. This includes the principles of paramagnetism and how they relate to the soil and to our health and well-being.

According to the basic principles of paramagnetism, igneous and volcanic rocks contain the highest paramagnetic qualities. In powdered or granular form they can be used as an "inorganic fertilizer" to increase both the quality and productivity of conventional food crops. In soils with high paramagnetic values, water from rainfall or irrigation becomes "energized" so it functions to keep plant fluids circulating at optimal levels, increases their ability to resist diseases and helps protect them from winter frost kills. It is also thought that paramagnetic minerals in the soil raise the vibrational vitality of crops so they can better repel insect pests.

According to Dr. Callahan, plants receive about 80 percent of their nutrition from the atmosphere. Most of this nutrition comes from carbon dioxide and water vapor, airborne nutrients in dust particles, and radiations from solar and cosmic energy. Atmospheric oxygen is also important for proper root growth and for those parts of the plant that are exposed to the atmosphere. As with volcanic rock and certain soils, oxygen itself is also highly paramagnetic and thus plays an important role in plant growth and vitality. Dr. Callahan maintains that most weeds are "green containers" for paramagnetic minerals that come from compost and manure, and can be beneficial if properly mulched and composted to recondition depleted soils. He observed that the healthiest crops grow in highly paramagnetic soils. Thus, the original vitality of depleted soils can be restored by adding pulverized igneous rock such as basalt or granite to the soil.

Through the use of a paramagnetic meter, which Dr. Callahan invented, farmers can avoid the use of chemical fertilizers and pesticides, by using scientific feedback from this instrument to guide them. Dr. Callahan estimates that on a worldwide basis approximately 60 to 70 percent of the original paramagnetic soil components have been depleted by conventional farming practices and soil erosion. He believes that natural soils should be teeming with living organisms such as bacteria and earthworms, and should also contain materials such as compost and paramagnetic minerals. He adds that a paramagnetic meter is critical for analyzing soils and for providing feedback to farmers who wish to bring soils back to their original paramagnetic states.

The paramagnetic forces inherent in igneous rocks serve to amplify extremely low frequency (ELF) waves in the atmosphere and also photon waves which lie within the visible and infrared spectrums. As with the human nervous system, plants are also photonic, as their life-processes involve photon transfer as a dynamic part of their metabolic cycles. Thus, the science of paramagnetism would seem to offer a time-tested and ecologically sustainable way to regenerate depleted soils and to produce fast-growing, nutritious organic superfoods (Callahan, 1994, 1995).

∞

In-Vitro Culture and Bioreactors: Food Production for the Future?

The Plant Biotechnology Lab in Disney Epcot Center in Orlando, Florida is one of the world's leading centers for future-food technologies. Researchers there have significantly advanced plant tissue culture technology to the point where entire plants can be regenerated from tiny leaf-tissue sections or cell cluster samples. To support this new technology, Epcot Lab has established its own "living library" which consists of over half a million tissue cultures which are maintained in sterile flasks and culture vessels.

Advanced tissue culture technology has already been applied to strawberries, pineapples, carrots, potatoes, peanuts and trees such as evergreens and redwoods (Zey, 1998). New breakthroughs in tissue culture technology are already powering similar transformations in the sciences of agriculture and forestry. This approach can also revolutionize the production of animal protein, as it offers fascinating possibilities for what has been referred to as "victimless meat production" – culturing animal tissue without animal slaughter.

(www.wendywolfson.com/yahoo_site_admin/assets/docs/steak.18691205.pdf)

∞

Spirulina: Blue-Green Algae that Transforms Solar Energy into Superfood

The microscopic blue-green algae *Spirulina* is one of the most concentrated food sources on the planet. It contains up to 71 percent protein, 10 percent carbohydrates, 7 percent fiber and 9 percent minerals. It is the highest source of Vitamin B-12 available, containing all nine essential amino acids plus nine non-essential ones. It is approximately 25 times richer in beta-carotene than carrots and has 56 times more iron than spinach or steak. *Spirulina* is rich in glycogen and higher in chlorophyll than either alfalfa or wheat grass. For these reasons NASA has selected it as a primary high-energy food source for astronauts in space.

Spirulina is very efficient in converting solar energy, carbon dioxide and water into a digestible high-energy natural food. Since the cell body of *Spirulina* is not encased in a mucus membrane (as with other algal species such as *Chlorella*) it is more digestible than most other forms of micro-algae, with a digestibility rate of about 95 per-cent when compared to 40-50 percent for *Chlorella*, 39 percent for soy beans, 18-20 percent for eggs and beef and 7 percent for rice. It is also the world's highest source of Vitamin B-12, containing some 255 mg per 100 g. (By comparison, beef liver has only 80 mg of B-12 per 100 g.). The taste of *Spirulina* is also considered superior to other algal species.

Since the individual cell bodies of *Spirulina* have a natural tendency to clump together into filaments at the surface of the water this makes it ideally suited for mechanical harvesting. Most of the *Spirulina* from commercial aquaculture operations throughout the world is spray-dried. It can also be drum-dried, which causes it to turn brown with a taste similar to Japanese nori.

When *Spirulina* is fed to certain species of fishes, it can accelerate their natural sexual maturity by as much as 60 percent. It also enhances their color and is thus prized by fancy carp breeders and trout farmers. It is a well-known fact that it is a healthy supplement for dogs and cats, as it increases the shine in their coats. *Spirulina* is the primary food for flamingos, being one of the factors responsible for their bright pink feathers. It occurs naturally in Mexico, Ethiopia, Australia, Kenya and in hot springs, where temperatures range from 89.6 to107.6 degrees F (32 to 42 degrees C).

Spirulina was first introduced to the general public in 1827 by German scientist Deurban, who discovered it growing in Lake Chad, Ethiopia where the salinity was so high that neither fish nor shellfish were able to survive. *Spirulina* has been a traditional food source for natives in the region who harvest it with baskets, dry it in the sun and store it in powdered form. It is then mixed with wheat, baked into bread with spices, consumed as a soup or made into a natural confection. Natural populations of this blue-green alga can still be found in Lakes Elementia, Rudolph and Nakuru in Kenya; Lakes Alanguardi and Circu in Ethiopia; Lake Buccacina in Peru and Lake Texcoco in Mexico, where it has been commercially cultured for several years (Maynard, 2009).

Sugar: Is it Really "Killing Us Softly With Its Sweet Song?"

In September of 2013 a surprising scientific report from the Credit Swiss Research Institute revealed the staggering negative impacts of sugar on the health of average Americans. The research revealed that about 30-40 percent of healthcare expenses go toward addressing issues which are tied directly to excess consumption of sugar in the diet. From an economic perspective the scientists found that our national addiction to sugar results in a staggering one trillion dollars each year! The report also determined that several health conditions stood out above the rest: coronary heart diseases, type 3 diabetes and metabolic syndrome.

One year earlier, in 2012, Dr. Sanjay Gupta appeared on CBS's "60 Minutes," where he highlighted the work of endocrinologist Dr. Robert Lustig, who garnered national attention with a lecture entitled, "Sugar: The Bitter Truth." This lecture had gone viral in 2009. Lustig's research investigated the connections between sugar consumption and poor health in Americans. In a series of 12 research papers in peer-reviewed journals he identified sugar as a major culprit in the epidemic of degenerative diseases that widely affects the American population. Lustig's data clearly indicated how excessive sugar consumption is a key factor in the development of many types of cancer, obesity, type 3 diabetes, high blood pressure and coronary disease. His extensive research led him to conclude that 75 percent of *all* diseases in

America are brought on by our American lifestyle. They are thus preventable. What is perhaps most surprising is that prior to the 60 Minute's program in 2012, no one in the scientific world had acknowledged that there was anything wrong with sugar. This research gives anyone a golden opportunity to improve their health and well-being by simply cutting down or eliminating sugar and sweets from their diet. It is also important to understand that the negative effects of sugar tend to be magnified in children, since their bodies and brains are still developing (Null, 2014).

A Future-Science Strategy for the Production of Superfoods

I would like to propose a new global program for the production of high-energy superfoods using selected applications of intensive aquaculture technology. This program would focus on the cost-effective production of superior nutritional food products. Futuristic food production technologies can effectively reduce food-growing space by factors of hundreds of times – or more. Suggested cultured food species include marine yeasts, phytoplankton, rotifers, *Spirulina* algae and small crustaceans such as krill. Ecologically speaking, intensive food production is far more efficient in terms of labor, water, space and energy requirements than with conventional livestock farming practices.

Since algae and planktonic organisms exist at the bottom of the biological food chain they are very energy-efficient to produce. These simple organisms offer many other advantages: They are small enough to be eaten whole and have no bones, feathers, urine, feces, or other biological wastes to pollute the environment. These characteristics significantly reduce their environmental impact when compared to conventional livestock, which require large amounts of land, water, fossil-fuel-based chemicals and supplemental foods – most often made from ocean fish products.

Advanced hydroponics, aeroponics and intensive aquaculture technologies can thus be integrated into commercial food production systems, which can cheaply and efficiently produce hormone- and pesticide-free superfoods. Using fertilizers which have been bioprocessed from human and agricultural wastes would create another win-win situation for getting rid of these polluting waste streams by providing fertilizer for hydroponic and aeroponics crops or to produce algae for aquaculture. Selective breeding and ethically responsible biotechnology could also be used to produce superfoods, which would be flavorful, organic, inexpensive, fast-growing, and nutritionally superior and could be tailored for any cultural diet.

To further expand the superfoods paradigm, floating offshore sea farms can be created which pump cold, nutrient-rich water from the ocean depths to the surface, where planktonic algae would use sunlight to convert these nutrients into food for larger zooplankton, crustaceans and fishes. Underwater lights could be used to concentrate plankton, which would attract increasingly larger fish for periodic harvesting. Such floating ocean farms could also help reduce the mounting pressures on our depleted commercial fisheries stocks. This would allow these critical natural resources the opportunity to return to their former sustainable levels.

MASTER KEY 20
Psychic Technology (Psy-Tech)

> "Learning to be aware of your psychic abilities
> can be like remembering something you already know.
> The process involves refining and deepening
> your responses to subtle information,
> rather than tapping into sources of knowledge.
> It is as though psychic abilities were sleeping quietly
> in the background of your mind,
> waiting for a nudge to awaken and open the lines of communication."
>
> *Russell Targ & Keith Harary –The Mind Race: Understanding Your Psychic Abilities*

Psychic Technology (Psy-Tech) involves the use of our innate psychic abilities as a tool for problem solving and moving projects ahead efficiently. Psy-tech also offers intriguing possibilities for developing new approaches to medicine, biology and human consciousness technology. History provides numerous examples of kings, presidents and military leaders who have relied heavily on the advice of their psychic advisors. In Russia scientific programs for development of military psy-tech have been actively pursued since the end of World War II, although most of this information remained classified until after the fall of the iron curtain. The United States, Japan, China and other major world powers are also fully aware of the strategic value of psychic technologies and have their own covert programs for using remote viewers to uncover hidden military facilities and weapons of mass destruction.

The future-science technology perspective highlights the potential uses of psy-tech to improve the quality of life for humans, to restore the global biosphere and to facilitate the evolution of human consciousness. In short, this approach would seek to research and develop new scientific applications for existing psychic techniques such as psi-microvision, remote viewing or future-vision to facilitate major advances in science, technology and world affairs.

Psi-Microvision: Amazing Psy-Tech Tool for Advancing Conventional Science and Technology

To date, physicists have defined over 200 subatomic particles using sophisticated experimental devices in the field of hi-energy physics. However, unknown to most conventional scientists is the astonishing fact that detailed observations of molecular, atomic and subatomic structures were carefully recorded over 100 years ago! An unusual form of psychic viewing (known at the time as "magnifying clairvoyance" or "micro-psi") has existed for thousands of years as hidden knowledge, known only to certain Eastern adepts. In its more recent format, instead of psychically magnifying the material to be studied, the technique involves making oneself "infinitely small" which allows atomic and sub-atomic particles to be studied and sketched in detail.

In 1895, two distinguished members of the Theosophical Society, C.W. Leadbeater and Annie Besant, embarked on an intensive micro-psi research program, which they pursued for 38 years. Beginning with simple elements Hydrogen, Nitrogen and Oxygen these psychic researchers described in detail all the known elements of their times – as well as a variety of organic and inorganic compounds.

The object to be studied was viewed by the psychic, who described what they "saw" to an assistant who produced detailed sketches. Atoms were depicted as highly structured bodies with definitive external structures which were subdivided into increasingly smaller spheroid, oval and conical configurations. Since the atoms under observation were vibrating at extremely high frequencies, the researchers developed psychokinetic techniques for slowing this motion down, thus enabling them to make accurate counts and observations (Leadbeater and Besant, 2011; Murphy, 2012).

In 1979, Cambridge University physicist Dr. Stephen Phillips developed an updated version of classical quark theory, which subdivided each quark into sub-quarks (omegons). Interestingly, the discrepancies between established physics theory and Dr. Phillips' new theory were bridged by a reinterpretation of Leadbeater and Besant's micro-psi data. Eighty-five years earlier they had claimed to actually "see" the atoms and subatomic particles exactly as described in Dr. Phillips' 1979 scientific paper which appeared in the journal, *Physics Letters* (Phillips, 1980).

It is my own personal conviction that psi-microvision represents a valuable analytical tool for implementation in virtually all sciences, since materials and chemical reactions can be studied by trained psychics at molecular, atomic and sub-atomic levels. Potential applications encompass a vast range of categories, including everything from medical sciences to aerospace technology. The implications for major advances in many fields of science would include huge savings of time and money as well as the undiscovered possibilities for refining many areas of science and technology.

In this same context it is important to note that my conviction with regard to the effectiveness and potential applications of micro-psi is based on thirty years of personal observation and experience with my former wife, Sharon Maynard, as well as other individuals who possess this amazing psychic ability and have successfully integrated it into their professional practices.

Remote Viewing as a Key Aspect of Psychic Technology

U.S. Military initially developed the basic protocols for remote viewing for obtaining information about distant or hidden targets using extra-sensory perception (ESP). In its basic format trained remote viewers were provided with a set of numerical coordinates written on a slip of paper. These numerical coordinates had been designated (unknown to the viewers) as relating to specific information such as "weapons of mass destruction. The remote viewer then provided their impressions and other information about an object, event, person or location that was hidden from physical view and most often in a distant location. The term *remote viewing* was coined later in the 1970's by parapsychology pioneers Russell Targ and Harold Puthoff at the Stanford Research Institute.

Remote viewing emerged into the public domain in the 1990's, following the declassification of documents related to the Stargate Project – a $20 million U.S. Government-sponsored research program initiated in 1975 to determine the potential military applications of psychic technologies. The program was terminated in 1995 after it allegedly failed to produce any useful intelligence information.

From a Future-Science Technology perspective, the basic remote viewing protocol can be upgraded so that trained individuals can remote view anyone or anything by simply focusing on a specific individual, location or objective. Theoretically, this advanced type of remote viewing could also be combined with psi-microvision to view, for example, the chemical reactions taking place inside shielded reactor containment vessels in a nuclear plant. In the future it may even be possible for trained viewers to use psychokinetic techniques (as with the Besant-Leadbeater experiments) to slow down or speed up certain chemical or physical reactions in hi-radiation areas where humans could not otherwise go. From my own experience with trained psychics the ability to remote view without numerical coordinates has been validated sufficiently since I have personally witnessed accurate results on numerous occasions. Once this improved type of remote viewing reaches a critical mass this type of psychic skill could become much more common. If so, it would certainly provide a definite evolutionary advantage for humans everywhere. The following references provide a comprehensive overview of remote viewing: (McMoneagle, 1993, 1998, 2000; Millay, 1999; Rifat, 2001). Certification courses with trained instructors are also available at the Farsight Institute (Farsight Institute, 2013).

Future-Memory: A Psychic Technology for Predicting the Future

Future Memory is a little-known aspect of psychic technology, which involves the ability to experience events from the future – before they actually occur. P. M. H. Atwater, author of *Future Memory,* developed this concept while studying the interactions of human consciousness with matter, energy and light. She determined that future-memory is apparently a natural phenomenon. Her work is based on hundreds of interviews in the U.S. and Canada.

According to Atwater's findings children under the age of five use this ability naturally. At the age of four most children spend more time in the future than in the present. She regards this early preoccupation with the future as a healthy activity and feels it is congruent with development of the temporal lobes of the brain, which occurs at this stage of childhood. She contends that the future-memory phenomenon is not simply a new form of psychic anomaly, but rather a manifestation of higher brain development, which establishes a bridge between the physical world and the higher self.

From the perspective of Future-Science Technology, future-memory would thus seem to be a psychic ability, which develops in our childhood years but has been stifled by over-emphasis on spoken language and electronic media. This natural ability *can* be recovered by simply acknowledging the fact that *it exists,* then reinforcing the conscious recognition of *when* and *where* these events occur during our daily activities. With practice, future-memory can become as natural as our senses of touch, taste, vision, hearing and smell.

From Atwater's research in near-death experiences and with individuals who experience future-memory routinely, she presents the following overview of typical future-memory experiences: 1) Often, a feeling of heat and exhilaration, and sometimes a ringing in the ears is experienced. 2) The physical senses are heightened and there is a cessation of motion. 3) There is a general feeling of expanding space. 4) A future scenario may manifest unexpectedly, although without training, it may be difficult to separate this experience from the present. 5) The future-scenario often ends abruptly, leaving the experiencer unsure as to whether it occurred in the present or future, although there is usually a feeling that "something unusual" has happened. 6) Emotional aftereffects help keep the event in the short-term memory, although the rational mind most often tends to set it aside or forget it. 7) Such a future-memory event is triggered when the event manifests in the physical, thus creating a sort of "déjà vu" experience. 8) Future-memory experiences are often associated with the feeling that a "divine gift" has been received (Atwater, 1999).

Future-memory is a new psy-tech method for the expanding the human intuitive senses. It has been developed to the point where some individuals are able to routinely use this expanded consciousness tool in their daily lives. According to Atwater, future-memory represents a brain-shift into a higher operational mode. If this shift can be *recognized* and *nurtured*, we can consciously learn to reinforce and develop the appropriate neural pathways as well as the expanded abilities associated with this future-memory sense. In her book, the author provides instructions for how to "jumpstart" this natural ability. She goes on to say, "Being able to live in the future in advance, and remember what one did, alleviates much of the stress and fear the unknown variables can cause. This advanced preparation enables the human psyche to negotiate the demands of sudden change more smoothly. The ability imparts an immense sense of confidence and peace in individuals, no matter what age, and often leads to frequent flow states, whether internally experienced, or as an aspect of how one's life can flow" (Atwater, 1996).

The infant science of "Future Memory" thus provides another example of a developing psychic technology. This new psychic science has endless possibilities when integrated into mainstream science. The advantages of evaluating future scenarios are obvious. Examples might include predicting the probability of severe weather events or asteroid strikes, or the relative success of any actions for resolving such devastating events. This approach could save valuable time and millions of dollars (Atwater, op. cit.).

Psychic Intuitives and Corporate Mystics in Business

Psychic intuitives and corporate mystics are people who have learned to apply their intuitive senses to various aspects of their professional activities. In their book, *The Corporate Mystic: A Guidebook for Visionaries with their Feet on the Ground*, authors Hendricks and Ludeman draw on over 25 years of experience in working with corporate intuitives. From their extensive research they contend, "Corporations are full of mystics. If you want to find a genuine mystic, you are more likely to find one in a boardroom than in a monastery or a cathedral." They contend: "…We have discovered that the very best kind of mystics – those who practice what they preach – can be found in the business world. We are now convinced that the qualities of these remarkable people and the principles they live by will be the guiding force for twenty-first-century business."

The authors maintain that a major key to success for corporate mystics is to stick to saying only what is true, and to follow through with total commitment and consistency. They conclude that most people simply cannot relax in an atmosphere of lies, distortion and hidden agendas. In other words, open honesty pays off, as it brings out the best in everyone involved in any business dealing. This sense of "practical integrity" is thus not just a noble idea, but also an effective key to success.

In their consulting work, the authors found that the corporate intuitives they studied had the following traits in common: 1) absolute honesty in their dealings with themselves and others, 2) a sense of fairness in their business activities, 3) a high level of self-awareness, 4) a focus on contributing to their business associates and to the world at large, 5) a focus on spirituality rather than religious dogma, 6) the ability to work efficiently (getting more done by doing less), 7) the motivation to bring out the best in themselves and in others, 8) a sense of seeing problems as opportunities, 9) the ability to be flexible to change, 10) a good sense of humor, 11) a strong sense of self-discipline, and 12) the ability to establish a reasonable balance between their visionary abilities and the real world. Other key attributes included taking full responsibility for their thoughts, words and actions. The authors thus concluded: "Intuition is a natural gift, something to which we all have access. The main reason it is so important is this: It is a clear sign that you are connecting with your inner spiritual guidance system. Intuition is a direct signal from your deepest self that you are navigating from your true center" (Hendricks and Ludeman, 1996).

Becoming a practical mystic and learning to use our intuitive abilities should thus be encouraged by parents and teachers from a child's earliest years. Intuition is a powerful force, which can be strengthened by regular practice. Thus, nearly anyone can develop this powerful intuitive ability, simply by becoming aware of *when* it is operating, then applying it in practical everyday applications. Greater public awareness can serve to reinforce the power of intuition and make it increasingly acceptable to global society. Based on their experience with over eight hundred business executives the following quote would seem to validate the importance of intuition as a key aspect of psychic technology: "Successful corporate leaders of the twenty-first century will be spiritual leaders. They will be comfortable with their own spirituality. They will know how to nurture spiritual development in others. The most successful leaders of today have already learned this secret. Corporate mystics know that an organization is a collective embodiment of spirit, the sum total of the spirits of the individuals who work there. Those who think spirituality has no place in business are selling themselves short" (Hendricks and Ludeman, op. cit.; Small, 1995).

Possible Future Applications of Psychic Technology

It is interesting to envision the creation of an international research program where new scientific applications of psychic technology can be explored and transformed into practical applications. To cite one example, this approach could be used to develop a "Global Consciousness Internet" for purposes of uplifting the human consciousness and reducing political and cultural conflicts. Such programs could also refine the various subtle energy technologies presently available, with the objective of developing new programs in areas of business, global ecology and future medicine.

Such new research program could also assist in integrating psychic technologies with quantum physics, chemistry, molecular biology and medicine. By combining psi-microvision with these sciences, entirely new breakthroughs would emerge in all areas of science and technology.

New applications of remote viewing technology can be developed to shift the focus from "military intelligence" to "natural intelligence." This would open new possibilities for scientists to gain insights into the nature of life itself and could provide a cost-effective means for the psychic exploration of the deep ocean trenches and outer space. This approach could save millions of dollars in research funds and help avoid many of the hazards of human space travel. In conclusion, military applications of remote viewing are thus only the "tip of the iceberg" when one considers its other potential applications.

MASTER KEY 21
Psychic Biology

> "The reality of the world
> lies in fields which interact with other fields of energy
> in dynamic chaos patterns
> that are always evolving to higher levels of complexity."
> *Valerie V. Hunt, 1989, Infinite Mind*

Physicists and molecular biologists have finally reached the point where, at a quantum level, the boundaries between hard science and consciousness science become blurred. Working within a quantum-thinking mode, it becomes theoretically possible to alter the very foundations of life itself – or even create matter from energy. Although it is possible to observe and quantify biological systems and reactions using hi-tech instrumentation, it is through consciousness technology that startling new insights and techniques for psychobiology are yet to be revealed.

Dark-Field Microscopy and Live Blood Analysis: An Amazing Medical Protocol Which Can Interface with Psychic Technology

Conventional bright-field microscopes use a powerful light source *under* the microscope stage to illuminate a specimen, which is placed on a glass slide. The light is concentrated by a condenser to highlight the specimen. Dark-field microscopy uses illumination *from the sides*, which reveals the condition of a patient's blood. The illumination source can use visible or ultraviolet light, or can be combined with phase microscopy – a polarization process that enhances the details in live cells. Bright-field microscopy, on the other hand, traditionally uses stained dead or frozen specimens which are standard procedures for medical diagnosis.

Darkfield microscopy has clear advantages since it can be used to view, photograph and videotape living specimens on a timeline. This provides solid documentation for scientific analysis and further research. Through this technology it is possible to get *instant* feedback to determine if a patient's red cells are healthy and separate, or arranged like stacks of poker chips – a condition called *rouleaux*. Equally important are the numbers and sizes of different types of white blood cells, as well as the presence or absence of yeast cells (*Candida*), which resemble small treelike bodies floating in the blood plasma. Certain parasites in the blood plasma (such as the spirochete protozoan responsible for Lyme disease) can also often be diagnosed. Thanks to improvements in the technology and refinements in diagnosis, dark-field live blood microscopy is relatively inexpensive when compared to conventional medical scanning devices. It also represents an underutilized tool for psychic medical researchers. A video explaining the basics of darkfield live blood analysis is available at: www.youtube.com/watch?v=Iu2_kZOg6ak

About ten years ago I attended a workshop in dark-field microscopy. We took two identical samples of a volunteer's blood and put them on separate depression slides with a cover glass to avoid evaporation. One sample served as the control. The second sample was used to demonstrate the effects of meditation on the blood condition of the volunteer. After a ten-minute healing meditation with the sample on the screen, the second slide was put up on the main screen alongside the control sample. Even for those not familiar with dark-field microscopy, the positive effects of this human conscious interface were evident. The red blood cells were no longer stacked like poker chips, but now floated free in the blood plasma – a more normal condition.

This demonstration provided a simple but scientifically valid example of the psychic power of healing. What is important to remember, however, is that dark-field live blood analysis also represents a major portal for psychic technology to become more closely integrated into future medical protocols. The inclusion of digital photography and real-time video documentation is another important factor, since results can be thus documented in a format that is acceptable for peer-reviewed scientific research papers. The following videos provide demonstrations of darkfield live blood analysis: (Seeger, 2007, 2014; Shepherd, 2011).

Although dark field live blood analysis is a valid technology, it does have its detractors. For example, medical wellness celebrity, Dr. Andrew Weil, has publically labeled darkfield live blood cell analysis as "completely bogus" (Wikipedia, 2014). Study the references provided and you can judge for yourself.

For those who wish to experience darkfield live blood analysis, practitioners are available nearly everywhere and the cost is relatively inexpensive when compared to conventional medical diagnostic techniques. One such experienced practitioner is Susan Hutchinson of Phoenix, Arizona. Susan has a solid background in different alternative healing modalities and is a professional teacher, author and TV talk-show personality. She was trained by darkfield microscopist, Dr. Rodney Ray, who has performed over 60,000 live blood cell analyses to date. Based on her lifelong experience with natural and psychic healing methods Susan has created what she calls her Master

Class, which is based on "the interface of quantum physics with mind power" and "the bridge connecting science with spirituality." Susan's latest book, *Divine Tune-Ups: Keys to Awakening the Soul*, is available in e-book format at: www.amazon.com

Recently Susan visited our research center in Sedona to demonstrate her amazing dark-field live blood analysis technology. Based on my previous experience with this technology, I feel that Susan has taken this real-time diagnostic tool to the next level. She uses live blood samples to focus in on dietary imbalances and is able to provide special colloidal minerals, probiotics and micronutrients to correct the imbalances in her patients. She validates her healing results by analyzing another blood sample from the patient after one or two weeks of using these dietary supplements. The reason I included Susan's work in the Psychic Biology section is that intuitive healer, Jerry Wills, was also present at these sessions. Jerry was, for the first time, able to "see" under the microscope and on the computer display what he normally "sees" using his own version of psi-microvision. This interaction between a skilled medical intuitive and the living blood sample heralds a new era for future-science medicine, since it provides a new tool for psychics to validate their work. This is thus a "perfect marriage" of ancient and modern medical technologies.

An Amazing Psychic Technology for the Genetic Transformation in Plants and Animals

Dr. Mahendra Kumar Trivedi, founder and president of India's Society for Divine life, has apparently been gifted with unique psychic abilities, which enable him to transform the genetic matrix of plants, seeds, bacteria, viruses, fungi and humans through a process he refers to as "consciousness intervention." Trivedi's achievements include the healing of thousands of people who were cured of chronic pain and other diseases. What is remarkable about his method, compared with other natural healers, is that his healings were apparently the result of his energies making significant changes in his patients' genetic and biochemical systems. Documented examples of Trivedi's healings include: alterations in genetic structure, the transformation of prostate and endometrial cancer cells into healthy cells, bacterial mutations which were passed on to subsequent generations and reducing the viral load in HIV and Hepatitis B and C. Dr. Trivedi believes these positive changes are caused by the effects of his energies on biological systems at atomic and molecular levels.

In experiments conducted by reputable universities in India, Dr. Trivedi was found to be able to permanently alter the genetic constitution in over 40 varieties of seeds. This resulted in significant improvement in crop yields and food value – all without the use of commercial fertilizers or pesticides. Experiments with scientists at various metallurgy laboratories determined that Dr. Trivedi's thought transmissions could apparently *alter* and *re-program* the atomic matrices of crystals and molecules, resulting in permanent changes in these substances. This research was validated using X-ray diffraction technology, electron spin-resonance spectroscopy, particle size analysis, Fourier transform infrared spectroscopy and ultraviolet spectroscopy.

Dr. Trivedi's ultimate goal is to bridge the gap between all branches of science that deal with living- and non-living matter. He envisions that a fuller understanding of his techniques will allow others to perform healings and influence the structure of matter. This, he contends, will usher in an entirely new dimension of science, which integrates consciousness, matter and energy into a single entity.

Within this same context, the fact that *one individual* has the psychic ability to effect positive changes in food crops and molecular compounds, suggests this ability can be taught to others. Thus, when a certain "critical mass" of trained individuals is reached, this unique ability will become a greater part of mainstream consciousness. If true, the implications for healing, the psychic enhancement of food crops and the transmutation of biological and non-biological materials offer a wealth of new possibilities for the future (Trivedi Foundation, 2010; Divine Life Foundation, 2010).

Fukushima Disaster Triggers Unprecedented Increase in Airline Pilot and Passenger Heart Attacks, Cancers, Radiation Illness Symptoms

In the face of the recent public concerns about eating tuna and salmon from the West Coasts of the U.S. and Canada, a new and more ominous threat has emerged – increasing levels of radiation in the atmosphere from the 2012 Fukushima Nuclear Plant disaster. Nuclear radiation expert and talk show host, Christina Consolo (aka "Radchick") has released the first ever FAQ (Frequently Asked Questions) on how the Fukushima radiation affects us when we travel by air. Christina's 25-page FAQ is one of the only information sources, which describes how this airborne radiation affects our families and us when we fly. Why is this important? The FAA and major airlines have yet to release any of this kind of information to the public!

The numerous revelations in Christina's FAQ and her Alfred Webre television interview include the following: 1) During one 2013 flight from the U.S. East Coast to Cancun, Mexico, Christina took periodic radiation measurements during the flight and found that she and her daughter were exposed to 1/10 of the FAA-mandated *yearly* radiation exposure limit. After the flight, while in Mexico, both mother and daughter suffered kidney failure and several passengers and crewmembers of that flight developed radiation-related illnesses. She also pointed out that TSA is now taking Geiger counters away from passengers as they board airliners. 2) Prior to the Fukushima disaster there were only six reported cases of pilots passing out in the entire history of commercial aviation. In her research Christina discovered that following the Fukushima disaster there was a marked increase in the number of radiation illness anomalies. These incidents included: heart attacks, pass-

ing out, health issues and unruly passenger reports (radiation is known to affect the executive function of the brain). This data was compiled by Christina from news reports and included in her FAQ – the first time this kind of information has been made public, since The National Transportation Safety Board does not presently document in-flight medical emergencies. 3) Since most U.S. celebrities are frequent flyers, Christina compiled an updated list of celebrities affected by radiation-related illness symptoms. This was in addition to the list she and independent scientist, Lauren Moret, had released in 2013. Examples among those celebs included: Selena Gomez, Rita Ora and Jennifer Lawrence (illness following long flights, requiring hospital visits) and Denise Richards, Angelina Jolie and Nicole Richie (gaunt appearance, requiring vitamin IV's). This new phenomenon is just beginning to surface at this writing, since it would appear to be only the tip of a much bigger iceberg as far as the Fukushima nuclear disaster is concerned. The following references provide an overview of this airborne radiation situation: FAQ: "Radchick: Fukushima Triggers Jump in Airline Pilot/Passenger Heart Attacks, Cancers, and Rad Symptoms."
(www.youtube.com/watch?v=EOtxx7zpyz0&feature=youtu.be);
"Radchick: Nuked in the Sky" (www.youtube.com/user/ichicax4) and, "Fukushima Falling Apart – Get Ready 2-14-2012"
(www.12160.info/forum/topics/fukushima-falling-apart-get-ready-2-14-2012) [Stay tuned, as more information is released and the "rest of the story" continues to be revealed].

MASTER KEY 22
After-Death Communication:
Removing the "Death Barrier"

> "Death is simply a shedding of the physical body
> like the butterfly shedding its cocoon.
> It is a transition to a higher state of consciousness
> where you continue to perceive and understand,
> to laugh and to be able to grow."
>
> *Elisabeth Kubler-Ross, M.D.*

Some Early Personal Encounters with Death

When I began elementary school I lived in Kennebunk, Maine. My best friend lived next door. His father was an undertaker. Believe it or not, his name was David Angel! During our activities we often played in the barn behind his house. Several of the rooms there functioned as his dad's mortuary. I remember being in the barn on one occasion when suddenly I spied a pair of bare feet sticking out from a mortuary table in one of those rooms. The sight of those bare feet haunted me periodically for quite a while.

Later, we moved to Farmington, Maine, where I attended elementary school as a fourth and fifth grade student. One of my friends, Taxi Davis, also had a dad who was an undertaker. I remember one evening we rode our bicycles over to his dad's funeral home. We got in through the back door and walked through a couple of rooms where coffins were displayed. The only light was from moonlight coming through the windows – very spooky. He opened the next door and turned on the lights. This revealed a room with several steel tables, which were used to drain the blood from corpses and replace it with embalming fluid. There before me, lying peacefully on one of the tables was a little old lady with her gray hair hanging halfway to the floor. Her naked body was partly covered by a sheet. While I remained in a state of shock, my friend stepped up to the table and said, "Want to see her nookie?" I was horrified and just wanted to get out of there. To this day the image of that little old lady is still emblazoned in my mind. I guess twelve-year-old-boys will be twelve-year-old boys! This was one experience I would never forget!

Some years later, while teaching comparative anatomy labs at the University of Miami I remember running occasional errands to the gross anatomy lab at the nearby Medical School, where medical students learned human anatomy by dissecting human bodies. Since dissecting preserved frogs or cats, and handling human bones, brains, hearts and other body parts was simply part of my job as a teaching assistan, I no longer found the spectacle of a partially dissected human body to be particularly shocking. Thinking back to my high school days and the death of my friend, Tony, I realized that even then, I had a deep knowing that the human body was only a vehicle for the humans I had known and loved, and that when an acquaintance or loved one passed away they would be able to leave the difficulties of living in the physical world, and go on to a place where they could experience peace, happiness and new beginnings.

How the Death of a Loved One Can Become a Golden Opportunity for Re-Inventing Ourselves

Death of a loved one, a child or family pet can be a mind-numbing and shattering experience. If approached with the proper attitude, however, this type of life-changing experience can provide a golden opportunity for personal growth, rebirth and transformation. Imagine how different our outlook on life would be if we replaced the concept of "dead and gone" with the concept of "death as a new beginning" – something which should be celebrated, since it frees the departed spirit from the discomforts of old age and the emotional burdens of life in the physical realm. Within this enlightened perspective death is really a rebirth into a new framework, which exists outside of the physical reality.

In 2007, I experienced the sudden death of my former wife and mission partner of over 30 years. To make a long story short, she suffered a heart attack and was transferred to a special cardiac hospital for emergency open-heart surgery. Following surgery, she never regained consciousness, and passed away quietly as soon as her life support systems were shut down. [I remember her telling me on more than one occasion "If you love something, you must be willing to set it free."]. This thought was still in my mind when I made the decision to shut off her life support systems, as I know she would have done the same for me had our positions been reversed. My first thoughts at that time were that my whole world was completely shattered. For the first time in my life, all previous plans for the future had become a blank page waiting to be written.

As a result of this tragic event, I found myself in an entirely new framework for my own survival and state of mind. A couple of weeks later, when I flew to Western Canada to attend my wife's memorial service, I discovered that she had advised a close friend there to seek the help of a psychic to resolve certain issues in her personal life. The friend contacted a professional psychic in the area, but had not made an appointment for herself. Near the time of the memorial service she called me and said this psychic had called her, and expressed with some urgency, that Sharon had been coming through to him. He told her he had messages for me and for our executive assistant who accompanied me. Although we had never met this individual previously, we called and arranged to meet with him at his home that same day. When we arrived, he greeted us at the door and ushered us into a room where he worked with his clients. We sat down at a table set for tea, not knowing what to expect.

Following a brief introduction, the psychic, Jesai Chantler, told us he was a professional psychic who specialized in the field of After-Death Communication. He went on to say he had never met Sharon in the physical, but had been going about his daily routine when he began receiving strong messages from her that he needed to make contact with us. He also explained how he had felt compelled [as if by some invisible force] to get the house perfectly clean, arrange the furniture in a specific way and even set four places for tea in the room where the meeting would take place…one for me and my assistant, one for himself and one for Sharon.

For the next few hours Jesai conveyed to us a series of messages from Sharon, as well as answers to questions we had about her sudden passing. This information provided us with a much-needed lift in spirit, so we could attend Sharon's memorial service and return to Arizona with a better sense of emotional resolution – and a measure of hope for the future. Back home in Arizona, I initiated a series of phone sessions with Jesai that continued over the next three years. During this period of my life, I realized that because of the spiritual work Sharon and I had done together, we had never accepted the concept that death was a final barrier to communication. This deep conviction compelled me to research the fields of after-death communication, as I was determined to carry on the goals we had established during our time together in the physical realm. I felt that if this information could be shared with others, it could help them resolve the emotional drama associated with death, by learning to better communicate with our loved ones who had passed over. This process was also helpful in resolving any karmic and personal issues which had occurred during our physical life together.

In addition to Jesai's expertise as an after-death counselor, he is a skilled psychic who once had his own radio show in British Columbia. He is also a certified Reiki Master and a practitioner of traditional Chinese medicine and acupuncture. As with most professional psychics he can work with clients in person or by phone. Additional information on Jesai can be found at: www.spirit-forces.com/id2.htm

Neurosurgeon Claims to Have Visited the Afterlife: Says, "Heaven is Real"

Dr. Eben Alexander has taught at Harvard Medical School and established a credible reputation as a neurosurgeon. Despite his Christian upbringing he claims he never had any special belief that an afterlife existed before undergoing a near-death experience, which changed his life forever.

In the fall of 2008, Dr. Alexander, who was suffering from a bout of meningitis, slipped into a coma for a week, during which time the neocortical part of his brain ceased to function. While in this comatose state he claims to have experienced a life-changing visit to the afterlife.

Dr. Alexander, who was featured in a cover story of *Newsweek* (October 15, 2012), relates his experience as follows: "According to current medical understanding of the brain and mind, there is absolutely no way that I could have experienced even a dim and limited consciousness during my time in the coma, much less the hyper-vivid and completely coherent odyssey I underwent" (Newsweek, op. cit.).

Dr. Alexander goes on to describe his impressions of heaven as follows: He relates how he found himself floating above clouds and watched "transparent, shimmering beings arching across the sky, leaving streamer-like lines behind them." He tells of being escorted by a female companion, and how he communicated with these heavenly beings in a way that transcended conventional language. He claims to have received messages from these beings which he paraphrased as follows: "You are loved and cherished dearly forever;" "You have nothing to fear," and "There is nothing you can do wrong." Dr. Alexander describes how he traveled to "an immense void, completely dark, infinite in size, yet also infinitely comforting." His impression was that this void was the home of God.

Following his recovery Dr. Alexander was at first reluctant to share this transcendental experience with his medical colleagues. Instead, sought out solace in his church. Later, he made the decision to take his story public and subsequently published a book about his experience: *Proof of Heaven: A Neurosurgeon's Journey into the Afterlife*. He comments as follows: "I'm still a doctor, and still a man of science every bit as much as I was before I had my experience. But on a deep level I'm very different from the person I was before, because I've caught a glimpse of this emerging picture of reality. And you can believe me when I when I tell you that it will be worth every bit of the work it will take us, and those who come after us, to get it right" (Alexander, 2012).

The more creatively we learn to express ourselves in the physical realm; the fuller will be our expression of the creative force in the afterlife. When we enter the spirit world our energy will be further expanded and accentuated. Thus, we should learn to regard our physical life as an incredible opportunity for growth and adventure. We should never let a single day pass without trying to move ahead by expanding our perception and personal expression of the creative force. Our physical existence could thus be compared to a sort of "lucid dream" which we all experience to enhance our universal perspective. What makes this so exciting is that we can learn to *change the dream* and thus experience growth and evolution to the fullest extent. Spiritual evolution is not so much like climbing a ladder as it is like fitting together the pieces of a puzzle. To "pass go" on our evolutionary pathways, we first have to experience the totality of the physical plane over countless lifetimes. During this process we go through a "school of hard knocks," where we are exposed to situations of negativity, and where new realizations present themselves in our consciousness with unexpected abruptness. This process is necessary for the higher self to develop more quickly.

By dealing with *who we really are* and *what obstacles we face*, we take responsibility for our actions and can move beyond the need to experience additional challenges and negativity. Once we come to the point when we totally accept that it is *we* who control our lives, a door swings open and our pace of evolution and awareness begins to accelerate. Without negativity being a significant part of our life curriculum, there would be no reason for the higher self to reincarnate.

The following resources provide a comprehensive overview of After-

Death Communication: (Anderson and Barone, 1999; Atwater, 1999, 2004, 2007; Bissler, Florino and Ruble, 2008; Bissler and Heiser, 2008; Botkin, 2005; Center for Spiritual Understanding, 2013; Goldberg, 1997; Guggenheim and Guggenheim, 1995; Harder, 1993; Martin and Romanowski, 1997; Orr, 1998; Weenolsen, 1996) and (Schwartz et al., 2002).

MASTER KEY 23
Channeling and Interdimensional Communication as Key Resources for Consciousness Development

> "Learning to work with spirit guides in various forms
> is a magnificent adventure.
> We open ourselves to great color and wonder.
> We can eliminate our primal fears of death and birth
> by connecting with those who have come and gone before us.
> We can experience the touch of loved ones beyond the physical,
> and learn to transform mourning into celebration."
>
> *Ted Andrews, 1994 – How to Meet and Work with Spirit Guides*

A "channel" is an individual who has developed the ability to set aside their own consciousness and allow another level of consciousness to come through – most often to the exclusion of their own conscious awareness.

From a historical perspective a "shamanic trance" is an experience associated with indigenous vision quests. Traditionally, a trance state was induced by hardship, isolation from other humans or psychotropic drugs. The objective of a vision quest was to gain wisdom and maturity, and to receive information from higher sources, which could then be shared with others. In a typical mediumistic trance (as with the oracles at Delphi) the oracles were trained to allow another entity to use their mind and body as a vehicle for transferring information from the higher realms.

"Conscious Channeling" As a Refined Version of Trance Channeling

With the more refined versions of channeling, the conscious mind simply steps aside, allowing the channeler to become an objective observer, but also remaining fully conscious of what is going on. By thus circumventing the ego-mind, it is possible for anyone to develop the capability to explore other dimensions, communicate with Spirit Guides and return with the information…with full memory and consciousness. The process is proactive in this format, since the channeler has full control of what is taking place. This type of conscious channeling opens a virtually limitless access to higher dimensional information resources. In practice, channeling sessions are normally recorded so clients can access this information at a later time. [Consciousness channelers liken their experience to "setting aside the conscious mind, which hovers near their shoulder as an objective observer."]

Channeling as a Valuable Para-Scientific Tool for Accessing Information and Guidance from the Higher Realms

The use of consciously controlled channeling to access information and guidance from the higher realms is an important key for evolution of the human consciousness. Channeling creates a communications link between the physical mind and the higher vibrational realms, where a collective consciousness exists that has often been referred to as *The Universal Mind*. When we channel, we can access this universal mind by connecting with our spirit guides, or our higher selves. These entities have the ability to *step down* their vibrations so they can answer questions and communicate from their own realms of refined consciousness. Access to this amazing resource might be compared to being allowed to access to a cosmic computer, since through channeling technology we can have our questions answered, solve problems and select the best possible future scenarios for any situation we encounter. Channeling thus offers the possibility for accessing all the ideas, knowledge and wisdom that have ever been or will be known. Although most of us are not aware of it, we already channel to a certain extent – especially when flashes of inspiration, guidance and creativity come to us during those special "a-ha moments."

When we put forth the firm intention to link up with our spirit guides, these guides can function as interpreters and teachers by showing us how to refine our abilities and help us to navigate these unknown realms. Spirit guides can also help us to learn to use the gifts and wisdom we gain to meet the challenges of daily life in the physical world. By learning to open our own channeling interface we no longer need to depend on other channelers or intermediaries to access this quantum information resource, since our prime objective in developing channeling abilities should be to first learn how to effectively *channel our higher selves* (Sanaya and Packer, 1987).

Personal Experiences with Channeling

Experienced trance channelers are adept at shifting their physical-based consciousness into an expanded state called a "trance." The basic format for modern trance channeling has been established for over 20 years by new age celebrities such as Lee Carroll and Darryl Anka who channel interdimensional guides "Kryon" and "Bashar" respectively. Both channelers have an extensive worldwide following. Lee Carroll has channeled Kryon for United Nations groups on several occasions (Kryon, 2010, 2012; Bashar, 2012).

Traditionally, the channeler sets aside their conscious mind. This is true of both Lee Carroll and Darryl Anka, who both introduce their presentations in their normal human personality then shift into a trance state where contact is established with the entity they channel. From my experience in meeting and observing both individuals at their channeling sessions, I found it remarkable how their voice patterns, body language and personality changed so dramatically at the moment when the connection with their spirit guides was initiated. This is especially true with Darryl Anka when he channels Bashar. This colorful entity speaks with a theatrical Shakespearian twist as he entertains his audiences with his outrageous and otherworldly sense of humor. In public appearances, sessions would begin with the channeler introducing himself, then explaining how he would slip into the trance state. Channeled presentations are usually followed up with sessions where audience members interact with the channeled entity by asking questions or offering comments. At the conclusion of each presentation, the channeler would slip back into his body and return to their normal human personality. In the case of Lee Carroll and Darryl Anka, neither individual claims to remember what takes place during these sessions.

Over the years, I have attended public sessions with both of these channelers. I also arranged for numerous personal sessions with others who would channel several entities or groups from the higher dimensions. I have attempted to view these sessions from a scientifically analytical perspective. In most cases I was impressed with the sincerity and professionalism of the channelers I worked with and have accumulated hundreds of pages of transcribed material from different higher sources. What I discovered was that, as is true with computers, it was critically important *to ask the right questions in the right ways*. Since internet data-mining and accessing information from the higher realms have many similarities, this is where I feel the "art of science" comes in, as when asking questions of a channeled entity it is first important to recognize the areas of interest and specialization of each entity or group you are working with and then ask questions which relate directly to their areas of expertise.

After attending several public channeling sessions, it became apparent that most of the audience questions to the channeled entity involved personal challenges or getting messages from a loved one who had passed away. Although this information was important from each individual's standpoint, using a cosmic entity to answer these questions is like water skiing behind an aircraft carrier! In my own channeling experience, I have found that these channeled entities and groups apparently enjoy communicating with someone who asks thought-provoking questions. I found this to be especially true of questions, which related to advanced science and technology, and how enlightened beings live and interact from their own perspective.

To cite one example of this principle, I attended one of Darryl Anka's channeling sessions with Bashar where I asked several questions about how to best access the information for writing this book. Bashar answered my questions and provided helpful answers. A couple weeks later I received an email from the videographer who was editing the filmed audience participation session. In the process of editing the sessions he was apparently struck by the way I asked the questions and how in line they were with his own on-camera questions for Bashar. After this initial contact we followed up with several phone and e-mail exchanges.

What is important here is to understand that different channeled entities come from vastly different cultures than our own. We need to express any questions in terms that make sense to the channeled entity and us as well. It has been my experience that if we address these beings in respectful and interesting ways they take equal delight in interacting with us.

In the past, channeling has received negative publicity – most often associated with psychics who channel entities from areas of potential negativity in the lower astral realms. They sometimes lose control over their bodies and thus control of *who* or *what* they are channeling. Numerous books and instructional courses have been developed which are available in print and DVD formats. This material is designed to teach us how to avoid negative pitfalls. The best way to avoid negative psychic inflows during this process is for the channeler to declare the strong intent before each session that any information coming from their sessions will be limited to that which is "for the highest good for all concerned."

Over the years, channeling has continued to be developed and refined.

As with other quantum-field skills it requires a combination of practice and discrimination. It is also important to cross-reference channeled information, by using other channelers, or by asking similar questions in different ways from these channeled sources. The ultimate objective of channeling should be to practice connecting with *our own* higher spirit guides. This way we can get the answers for ourselves *without* any intermediary. With practice, information, answers to critical questions and inspiration will begin to come to us automatically. Another key benefit of channeling lies in the fact that many of us fail to understand that it is the *energy* being transmitted from higher channeled sources that is *far more important* than the words alone. Awareness of this principle can thus make these sessions more informative and productive.

I have worked mainly with skilled psychics who are able to channel *consciously*. In other words they are able to retain an awareness of their own personal consciousness, as well as a full awareness of information imparted by the channeled entity. I thus feel that conscious channeling is a more evolved form of channeling, and that once a critical number of individuals are able to do this, it will become the preferred technique. For additional information on channeling the following websites are available (Sonora, 2013; Rachelle, 2013). A comprehensive book on channeling has been published by Jon Klimo, senior faculty member of the Rosebridge Graduate School of Integrative Psychology in Concord, California and Director of Doctoral Programs in Parapsychology. Through Jon's efforts over the years, channeling is on its way to becoming a respected academic discipline (Klimo, 1998).

Why Should Entities in the Higher Dimensions Ever Wish to Communicate with Humans?

In channeling sessions with different entities I felt compelled to ask the following question, "Why would evolved beings like you want to waste your time helping us, since from your perspective, interacting with humans must seem like a human talking with a mouse?" Their answers could be summarized as follows: "We do this, because we are curious to learn about and experience your culture, and also because we care about you and enjoy your company. We do this because it allows us to experience the mutual joy and creative spirit of your world. This is our greatest delight." Most of these higher entities and councils have strict codes of non-intervention, but *will* occasionally intervene and provide help under special circumstances – especially when it is requested. These higher beings all have in common a refined sense of humor. Regarding our earthly problems they had this to say: "You are the ones you have been waiting for" and that they had been waiting patiently for us to only put forth a specific request.

MASTER KEY 24
Exopolitics: Will Humans Ever Become Members of the Extraterrestrial Community?

> "Human understanding of the extraterrestrial phenomenon
> has evolved gradually over the past half-century.
> At first, there were arguments about whether the UFO-ET phenomenon
> was real or not.
> Next, the discussion moved to the exploration of the
> nature and the purpose of the phenomenon.
> Today, a conversation is beginning about initiating conscious human interaction
> with the life forms – the expressions of Nature – that we are calling extraterrestrials."
> *Michael Mannion – Project Mindshift:*
> *The Re-Education of the American Public Concerning Extraterrestrial Life*

The following statement was made by retired U.S. Air Force intelligence officer, Robert Dean, an individual who I have had the opportunity to meet in person and learn about his experiences with UFO's and different types of extraterrestrial beings. In the course of our conversation, Bob unequivocally stated that on several different occasions he had participated in meetings with leaders from the U.S. Military, government officials and different types of extraterrestrial beings. From my personal perspective I consider Bob Dean be a natural leader and an individual of high integrity who has the courage to reveal the truth as he has experienced it. His convictions about the reality of the ET presence in government affairs are expressed in his own words: "It has been 40 years now, since I first became aware of the extraterrestrial presence on Planet Earth. Since that experience, my life has never been quite the same. I was to learn that the human race has had, and continues to have, an intimate relationship with several incredibly advanced intelligent races from other planets, solar systems and star systems within our galaxy – and that this relationship has been underway for several thousand years.

These star-traveling civilizations are as far beyond humans on Planet Earth as modern America is beyond the head hunting tribes of New Guinea. This is primarily why disclosure has not taken place – and why disclosure is not contemplated by the unacknowledged US Government agencies that oversee this great secret. You see, we cannot open Pandora's Box just a little bit. Once we open it, nothing will be the same. A major new paradigm will come crashing in, and our old world will come crashing down. Religion, Society, Politics – all will be utterly changed forever. Obviously, this is what the world's governments fear.

The final reality is that the story must be told. Exopolitics is a logical, rational, and scholarly attempt to clarify and present to the world the structure of an existing reality that can become a valuable tool in educating and expanding the human consciousness. If we ever mature as a race, we must recognize our extended family, and reach out to them with courage and fellowship. Exopolitics can show us the way" (Dean, 2013).

About four years ago a professional colleague (David Sereda) suggested I call Boyd Bushman, an aerospace engineer who he thought might be willing to share some interesting information. Senior Research Engineer, Boyd Bushman, has worked for Lockheed Martin, Texas Instruments and Hughes Aircraft. He is regarded as one of the inventors of the Stinger Missile. In a series of on-camera interviews Boyd reveals how Lockheed Martin has pursued antigravity research (using magnetic fields to manipulate gravity). He also explains how, in his research with Lockheed Martin, he proved that magnetic fields affect Earth's gravitational field, so solid bodies *do not* fall with the same acceleration – a result that contradicts Galileo's famous experiments at the Leaning Tower of Pisa.

One day I decided to call Boyd – not really knowing what to expect. He answered the phone and I introduced myself. As we began to talk, Boyd began sharing some of his experiences in skunk works environments (Lockheed Martin is famous for their covert-ops "skunk works," a unique think-and-do tank, where many breakthrough military technologies have been developed). As a dyed-in-the-wool research scientist, I have a burning curiosity when it comes to dialoguing with this type of brilliant scientist. What I *am* most interested in is engaging in scientific dialogue with individuals who have high levels of integrity and are willing to share information tidbits about their work. When these top-level scientists understand you are not simply trying to "pick their brains," but instead wish to engage in a meaningful scientific discussion, they relax a bit and will usually share some of their "war stories." In my experience, most brilliant scientists are proud of their accomplishments, and understandably are eager to share them with others of like mind. This was true with my initial conversation with Boyd. We talked for nearly an hour.

In the ensuing conversation, Boyd revealed some off-the-record information, which I am not at liberty to share. What Boyd *did* tell me, however, validated the fact that some of the companies he had worked for *did* have alien artifacts in their possession. He also informed me that some of the advanced weapons technologies he had worked with far exceeded the laws of conventional physics. Boyd is a brilliant scientist and a caring individual of high integrity. To learn more about Boyd's work, his video interviews can be found at: www.youtube.com/watch?v=wT0uMo_TmiU

Information on Lockheed Martin's skunk works is available at: www.lockheedmartin.com/us/aeronautics/skunkworks.html

In closing this section I would add the deathbed confession of former Lockheed CEO, Ben Rich. Ben, known as "the father of the Stealth fighter-bomber," revealed the following information before passing away in January of 1995: "We already have the means to travel among the stars, but these technologies are locked up in black projects, and it would take an act of God to ever get them out to benefit humanity. Anything you can imagine we already know how to do." He continued, "We now have the technology to take ET home. No, it won't take someone's lifetime to do it. There is an error in the [Einstein's] equations. We know what it is. We now have the capability to travel to the stars. First, you have to understand that we will not get to the stars using chemical propulsion. We have to design a new propulsion technology." Ben apparently also knew that the crashed saucer(s) at Roswell, New Mexico influenced the designs of Testor's model of the Roswell extraterrestrial UFO, as well as the designs of top-secret U.S. military aircraft (Godlike Productions, 2014).

Exopolitics: A New Protocol for Human and Planetary Evolution

Historically we have been taught to believe that intelligent life was limited to Earth. Exopolitics presents a new paradigm, which has the potential to evolve into a new worldview. Within a new exopolitical framework, Earth would appear to be an isolated planet in a well-populated universe. This universal community apparently consists of many highly evolved, technologically advanced and consciously evolving civilizations. This constitutes a precious potential resource for stabilizing our socio-economic systems, cleaning up the environment; managing our population and helping us upgrade our human lifestyles. Hopefully, at some point we may be judged mature enough to join the Universal Community. Although it is well known that this universal community contains both positive and negative factions, the foundation for the more advanced galactic civilizations is one of cooperation, service to others and cultural and consciousness enlightenment.

Exopolitics and the UFO Phenomenon

> "All truth passes through three stages.
> First, it is ridiculed,
> second, it is violently opposed,
> and third, it is accepted as self-evident."
>
> *Albert Einstein*

The wealth of documentation containing reliable credentialed witness accounts of encounters with ET's and UFO's has reached the point where anyone who studies the records of UFO sightings, abduction experiences, cattle mutilations, etc. cannot fail to realize the supporting facts are impossible to deny. This material includes detailed interviews, official reports, photos and videos, and books by military and commercial pilots. The evidence is not only overwhelming, but comes from nearly every country in the world.

For those who are still skeptical, I would suggest cross-referencing information sources before questioning them in detail, as it has become apparent to dedicated scientists, journalists and literary researchers that *creating disinformation* has been an easy way for authorities to avoid provocative questions and to squelch unwanted media coverage. This "disinformation syndrome" has served to hinder and discredit investigators of integrity, even though they have exercised due diligence in their research, books and papers.

To make matters worse, many institutional scientists have proven to be so deeply entrenched in their own closed mindsets that they tend to resist *anything* that threatens their reputation with their scientific colleagues or funding sources. Thus, certain otherwise professionally respected individuals have not only refused to be open to the possibilities of the UFO phenomenon, but have irrationally spoken out against UFO's and advanced technologies like zero-point energy and cold fusion (LENR). This type of shameful behavior initially destroyed the credibility of cold fusion researchers Stanley Pons and Martin Fleischmann who, in 1989, were raked over the coals by colleagues because their results were not "consistently reproducible" by other scientists. It later became clear that the source of the inconsistencies (unknown to the researchers at the time) was due to the purity of the experimental electrodes and inconsistencies in their measurement methodology. Interestingly, researchers working in their spare time at the U.S. Office of Naval Research eventually validated their work (Krivit, 2007).

UFO's: The Overwhelming Evidence

Dr. James E. McDonald, Senior Scientist at the University of Arizona's Institute of Atmospheric Physics states: "From time to time, in the history of science, situations have arisen in which a problem of ultimately enormous importance went begging for adequate attention simply because that problem appeared to involve phenomena so far outside the current bounds of scientific knowledge that it was not even regarded as a legitimate subject of serious scientific concern. This is precisely the situation in which the UFO problem now lies. One of the principal results of my own intensive studies of the UFO enigma is this: I have become convinced that the scientific community, not only in this country but throughout the world, has been casually ignoring as nonsense a matter of extraordinary importance" (Hastings, 2008).

The sheer volume of UFO-related literature is overwhelming, so a partial review of existing literature is in order. For example, Leslie Kean's book, *UFO's: Generals, Pilots, and Government Officials Go on the Record*, provides documentation from around the world which include the following provocative incidents:

Portugal – "In 1982, Portuguese Air Force pilot Julio Guerra looked out his cockpit window down to the ground below, and saw a low-flying metallic disc. Suddenly it bolted up toward him at high speed. During a lengthy series of events, the object demonstrated a harrowing variety of maneuvers close to Guerra's small plane. This was also witnessed by two other Air Force pilots called to the scene."

Chicago, IL – "On November 7, 2006 something unimaginable happened at Chicago's O'Hare Airport during the routine afternoon rush hour. For about five minutes a disc-shaped object hovered quietly over the United Air Lines terminal and then cut a sharp hole in the cloudbank above as it zoomed off. Hardly anyone heard about it until the story broke on the front page of the *Chicago Tribune* on January 1, 2007 – almost two months later. This triggered a flurry of national coverage on CNN, MSNBC, and other networks. With over a million hits, the *Tribune* story quickly achieved the status of being the most-read piece in the entire history of the newspaper's website."

Iran – On the evening of September 18, 1976, the citizens of Tehran were frightened by an unknown object with flashing lights, which erratically changed positions. The Iranian Air Force was alerted and fighter jets were scrambled to investigate the object. According to the flight crew briefings, the object was flashing very rapidly with red, green and orange-blue lights, which were so intense that the body of the craft was not visible. The pilot described his experience as follows: "The lights formed a diamond shape." He then related the events that followed. "I approached, and I got close to it, maybe seventy miles or so in a climb situation. All of a sudden it jumped about 10 degrees to the right. In an instant! Ten degrees…and then again it jumped 10 degrees, and then again." The fighter jet was finally able to get close enough to the craft to fire, and got the targeting radar locked on. However, when the pilot attempted to fire, all instrumentation blanked out, and his communications became garbled, and then lost."

Peru – On April 11, 1980 an object resembling a weather balloon appeared in restricted air space above La Joya Air Force Base. Air Force fighters were scrambled. The details as related to Commander Oscar Santa Maria Huertas (now retired) are as follows: "I began closing in until I had it in perfect sight. I locked in on the target and was ready to shoot." As the chase continued he stated further, "I got as close as 300 feet (100 meters) from it. I was startled to see that the "balloon" was not a balloon at all. It was an object that measured about 35 feet (10 meters) in diameter with a shiny dome on top that was cream-colored, similar to a light bulb cut in half. The bottom was a wider circular base, a silver color, and looked like some kind of metal. It lacked all the typical components of aircraft. It had no wings, propulsion jets, exhausts, windows, antennae and so forth. It had no visible propulsion system."

In the words of the book's author, Kean, "…I am convinced that UFOs exist and cannot remain unacknowledged by governments. The phenomena are evident in all parts of the world and no efforts in their study should be neglected. Toward this end, international cooperation is vital to generate standards for protocols and policies for data analysis." She continues; "Personally, I believe that the UFO phenomenon is the most interesting of all the many phenomena affecting our planet, and it is one that totally defies logical explanation. As of now it seems beyond our ability to comprehend. But new cases continue to be documented by pilots, air traffic controllers, air operations staff at the world's airports and many others with the proper training to determine whether a flying object is something unusual. Even though the true origin of these UFO's remains unknown, they do affect aviation everywhere, and this must be addressed. Eventually, I believe, we will be able to determine the real nature of this phenomenon if the scientific method is applied" (Kean, 2010).

Efforts to get the U.S. Government to officially acknowledge the existence of UFO's and to publically disclose classified information have been ongoing at the White House level for several years. For example, a petition with the required 25,000 signatures was submitted to the White House recently, but no official statement has yet been forthcoming...this despite efforts of President Jimmy Carter, President Kennedy, US Representative Dennis Kucinich, Barry Goldwater, Arizona Governor Fife Symington and others. Despite the efforts of the powers that be to restrict UFO disclosure in the United States, many other countries have already released formerly classified UFO documents. These nations include: France, England and the UK, Mexico, Chile, Peru, Uruguay, Brazil and Iran (Kean, 2010, op. cit.).

In the U.S. there has been significant progress in pressuring our government for open UFO/ET disclosure. Key proponents for dissemination of classified information include Stephen Basset, Executive Director of the Paradigm Research Group in Bethesda, Maryland. Steve presents himself as "an official lobbyist for disclosure" by the U.S. Government and Military organizations. Steve has organized numerous conferences, set up worldwide disclosure websites and has participated in television and radio interviews throughout the world. He has also spearheaded several press conferences at the National Press Club in Washington, DC, which involved top ranking U.S. and foreign military personnel. He has been responsible for sending petitions to President Obama and the White House, which request disclosure and the banning of nuclear weapons (Paradigm Research Group, 2013).

In an effort to educate government leaders to be more comfortable with the ET/UFO phenomena, Arcos Cielos Research Center, with support from Director Robert Miles, produced a special congressional edition of Robert's landmark documentary, *Fastwalkers: They Are Here: UFO & Alien Disclosure*. With the cooperation of Steve Bassett and his PRG associates, 350 copies of this video were created and hand delivered to every congressional office on Capitol Hill. Another 300 copies were similarly delivered to the offices of Canadian Parliament in Victoria, British Columbia. Copies of this provocative film are available at Robert's website at: www.fastwalkers.com (Miles, 2007).

Another individual who has been instrumental in bringing the ET/UFO phenomenon into the public arena is Steven Greer, MD, who heads up The Disclosure Project (www.DisclosureProject.org) and The Center for the Study of Extraterrestrial Intelligence (www.CSETI.org).

Dr. Greer is the author of several books, which include: *Extraterrestrial Contact: The Evidence and Implications, Hidden Truth: Forbidden Knowledge* and *Contact: Countdown to Transformation* (Greer, 1999, 2009, 2011). Dr. Greer has addressed radio and television audiences worldwide where he has shared his own personal disclosure experiences from his early childhood on. He has attended high-level meetings with over 450 military and government insiders and whistle blowers, and has participated in briefings with senior government officials, who include former CIA Director, R. James Woolsey, members of the U.S. Senate and senior United Nations officials.

One recent disclosure event involves one of Dr. Greer's key witnesses for the hearing with U.S. Congressmen and Senators. This remarkable video features former Space Technology Transfer Consultant and whiz kid rocket scientist, David Adair, who speaks out for disclosure for the first time – after remaining silent for many years. David tells the story of how in 1971 at age 17 he constructed an advanced electromagnetic containment plasma fusion rocket, which exceeded any known military technology. He was invited by Air Force brass to test his rocket at the White Sands Missile Range in New Mexico and was subsequently taken to Area 51 near Groom Lake, Nevada. There he met Air Force General, Curtis Lemay, and was shown an electromagnetic plasma fusion rocket of advanced design which he describes as being the size of a school bus. Upon examining this device he determined that it was an example of what he later termed, "advanced symbiotic technology," since the unit lacked any visible wiring (as with his own rocket engine). He further determined that the engine, which had been damaged and was unworkable, was "alive" in the sense that this advanced technology interfaced with a pilot's brain, being designed to function as a "consciousness interface." He deduced that this engine was alien technology that was essentially light-years ahead of anything that existed on Earth (UFOTV, 2010).

UFO Incursions into U.S. and Foreign Nuclear Military Bases

> One of the principal results
> of my own intensive study of the UFO enigma is this:
> I have become convinced that the scientific community,
> not only in this country, but throughout the world,
> has been casually ignoring as nonsense
> a matter of extraordinary importance.
>
> *Dr. James E. McDonald, Senior Physicist, Institute of Atmospheric Physics, University of Arizona*

Author Leslie Kean also points out that from 1945 to 1998 records show that a total of over 2400 nuclear devices have been detonated (543 atmospheric tests and 1,876 underground). She compares this data with 150 visual and radar UFO sightings. The similarities created by superimposing the data suggest that this well-documented presence of UFO activity is directly related to strategic nuclear activity worldwide (Kean, 2010, op. cit.).

The UFO-Nuclear Site connection is also covered extensively by veteran researcher, Robert Hastings, who has investigated nuclear weapons-related UFO incidents for over thirty years. Hastings interviewed hundreds of U.S. Air Force personnel who witnessed extraordinary UFO encounters at nuclear weapons sites around the world. Examples are as follows: In March of 1967 two noteworthy events occurred outside Malmstrom Air Force Base, Montana. According to six former, or retired, U.S. Air Force officers, "UFOs apparently disrupted the functioning of all 10 Minuteman 1 (ICBM) missiles at Echo Flight, on March 16th, and essentially repeated the feat at Oscar

Flight – around March 24th – when at least six to eight missiles were simultaneously shut down." One of the witnesses Hastings later interviewed was former U.S. Air Force Captain, Robert Salas, who at the time was Deputy Missile Combat Crew Commander. Salas, who was 65 feet underground in a launch command capsule, told the author he had received an urgent call from his Site Security Controller that several security guards were reporting strange lights maneuvering in the sky near the launch facility. Salas comments that it was "very rare" for even a single Minuteman missile to malfunction, and the fact that so many of them had suddenly switched to a "no go" status stunned Salas as well as his fellow launch officer. "We couldn't believe it." Salas said. He added, "As I recall, most of the failures were related to the guidance and control systems." According to Hastings the U.S. Air Force has remained silent on the incidents surrounding the Echo and Oscar Flight shutdowns. Apparently these two major disruptions (each lasting a day or more) created great concern for the local Air Force brass as well as for the highest levels of the Strategic Air Command. Says Hastings, "In short, the two closely-spaced incidents had impacted U.S. national security in the most fundamental manner."(Hastings, 2008).

Astronaut Edgar Mitchell's Comments on UFOs and ETs

Astronaut Edgar Mitchell is best known as the pilot for the Apollo 14 lunar module. He was the sixth person to walk on the moon and spent a total of 14 hours on the lunar surface. He was presented the Presidential Medal of Freedom by President Nixon in 1970 and left NASA in 1972. Ed's other interests include consciousness research and paranormal phenomena. On his return trip to Earth, Ed claims to have undergone a powerful Samadhi experience, and relates how he conducted ESP experiments with friends on Earth. In 1973 he founded The Institute of Noetic Sciences (IONS), a non-profit organization created to research areas overlooked or ignored by mainstream science. Ed's book, *The Way of the Explorer*, discusses his journeys into mysticism and space (Mitchell, 2008). The event that sparked Ed's interest in paranormal phenomena occurred during his teenage years when he met another teenager from Vancouver, Canada, then known as "Adam Dreamhealer." According to Ed, Adam helped him heal kidney cancer at a distance. "I had sonogram and MRI scans that were consistent with renal carcinoma." Following a series of healing sessions with Adam, Mitchell stated that, "...the irregularity was gone and we haven't seen it since" (Wikipedia, 2013, Edgar Mitchell).

Ed has gone on record, stating that he is "... 90 percent sure that many of the thousands of unidentified flying objects, or UFOs, recorded since the 1940's belong to visitors from other planets." He goes on to say that UFO's and ET's have been, "the subject of disinformation in order to deflect attention and to create confusion so the truth doesn't come out." In a 1996 interview on *Dateline NBC*, Ed discussed meetings with officials from three different countries who claimed to have had personal encounters with extraterrestrials, and stated that, in his opinion, the evidence for such alien contact was "very strong," and had been "classified" by governments, who were attempting to cover up visitations and the existence of alien beings' bodies in places such as Roswell, New Mexico. He went on to say that UFOs had provided some "sonic engineering secrets" that were apparently helpful to the U.S. government (Dateline NBC, 1996).

In a 2004 interview with the *St. Petersburg Times*, Ed suggested that a "cabal of insiders" in the U.S. government were studying recovered alien bodies, and suggested that this group had stopped briefing U.S. presidents after John F. Kennedy. He said, "We all know that UFO's are real; now the question is where they come from"(Moore, 2004).

In a 2008 interview with Nick Margerrison of Kerrang Radio Ed claimed the UFO crash in Roswell, New Mexico was real. He expressed his opinion that aliens have contacted humans on several occasions, but that governments have hidden the truth for 60 years. Ed stated, "I happen to have been privileged enough to be in on the fact that we've been visited on this planet, and the UFO phenomenon is real" (Kerrang Radio, 2008).

In a July 25, 2008 *Fox News* interview, Dr. Mitchell stated that his comments did not involve NASA, but quoted unnamed sources at Roswell (since deceased) who had informed him in confidence that the Roswell incident *did* involve an alien craft. He also claimed to have subsequently received confirmation of this information from an unnamed intelligence officer at the Pentagon (Fox News, 2008).

Exopolitics: A Wild-Card Solution to Our "Global Dilemma"

Recognizing that our global community is only part of a greater cosmic community provides a golden opportunity for solving these problems through the assistance, intervention and meaningful intercultural interchange with our extraterrestrial counterparts. Exopolitics thus provides us with a new universal paradigm, which can evolve into an entirely new global worldview. The old Twentieth Century paradigm posited that intelligent life, as we know it was limited to Planet Earth. Within the new exopolitical framework, Earth would now appear to be an isolated planet in the midst of a well-populated universe. This universal society apparently consists of many highly evolved and advanced civilizations, which offer the promise of helping us advance our consciousness, clean up our polluted planet, stabilize our social and economic systems – and ultimately be judged mature enough to join "The Universal Community." Although it is well known that this universal community contains factions with both positive and negative agendas, the foundations for most of these galactic civilizations are apparently based on service to others.

In the words of Canadian Alfred Webre, one of the founding fathers of exopolitics, "The exopolitics model holds that Earth is presently isolated from interaction with organized intergalactic civilizations, because it is under intentional quarantine by a rational, structured universal society. There are signs around us of a universe initiative to reintegrate Earth into interplanetary society. It is possible that Earth may be permitted to rejoin this universe society, under certain conditions at a future time." Webre feels that evidence from the records of past civilizations as well as more recent information would suggest that planets are "grown like gardens," and are therefore the products of the conscious intervention by advanced scientific beings that he refers to as "the life-technologists of universal society" (Webre, 2005).

Exopolitics embodies quantum-field thinking, in that many advanced extraterrestrial civilizations apparently exist in other dimensions which function at higher frequencies than our own. Accordingly, these advanced beings have full access to and knowledge of our thought forms and activities, as well as those of the countless other civilizations that populate interstellar space. Within this new exopolitical model, planets like Earth are members of a larger universal federation that operates under a complex structure of universal laws (Exopolitics, 2013).

Our future-thinking leaders have the opportunity to create bridges in understanding between humans and our ET allies. Considering the frequency shifts that Earth is presently going through, our genetic structure and interdimensional vision is apparently being adjusted so we can function as vanguards for the evolution of human society. These are the "gifts" we have earned for our dedication and diligence in seeking the truth, working to clear past emotional blockages and pursuing the expansion of our consciousness. By acknowledging these gifts, and by opening ourselves to work in the quantum field, we can continue to receive the light-encoded information packets downloaded to us from these higher information sources in accord with what our mental, emotional and physical bodies can handle at any given time. For this process to be effective it is important to stay focused and grounded into the physical reality, as we learn to apply these new tools in our personal and professional lives.

For those wishing to learn more about Exopolitics, the following documentary videos are suggested: *Fastwalkers: They Are Here: UFO & Alien Disclosure* (Miles, 2007); *UFO: The Greatest Story Ever Denied* (Escamilla, 2006); *The Disclosure Dialogues: A Historical Achievement on Five DVD's* (James and Stein, 2011) and *Alien Viruses: Crashed UFOs, MJ-12, & Biowarfare* (Wood and Redfern, 2013).

MASTER KEY 25
Quantum Leapfrogging: A Pathway to Transformation and Enlightened Evolution

> "A transformer creates the 'new' through that which already is,
> a reformer seeks to destroy that which already is,
> in hopes that something better will take its place.
> The reformer's work is ceaseless, never-ending, never satisfied.
> The work of the transformer is always complete and perfect within itself;
> It is always at peace as the energy streams through it out into the world...
> to heal, transform, energize and uplift.
> Allow yourself to be the transformer which you truly are."
>
> *Alexis Edwards, 1971 – Guidelines*

Quantum Leapfrogging is a term I use to describe the process of interacting with the universal information field so we can move forward in a series of quantum leaps. It works this way: Instead of moving passively into the future in normal linear fashion, it is possible to "leapfrog" into new creative mindsets and ways of thinking by learning to interact with the universal quantum field. With practice, this kind of quantum-leaping phenomenon will become as easy and unconscious as the act of breathing.

Moving into the Future Within the Field of Self-Creation

As we move through each life experience in our personal and professional lives, the lessons and emotional episodes become part of our accumulated wisdom bank. In the process of self-creation, one of the most powerful tools we possess is the quality of *desire*. This valuable tool can serve as a "prime driver" for creative success if we allow it the space to do so. It is thus important to remember that events and situations, which seem opposed to our core objectives, can be viewed as opportunities to test the "truth" of each given situation. True transformation is therefore not about trying to be different and pushing for answers, but in actually *living* and *becoming* the answer itself. It is also important to realize that once a certain thing or principle has been manifested for the first time by one of us, it will be much easier for it to be manifested by others (i.e. The Hundredth Monkey Effect). This same principle can just as easily be applied to our professional lives, as *no one* can ultimately deny the principle of *success*. Success is contagious; it can be caught and will then be emulated by others.

Transformers as "Catalytic Agents" for Raising the Global Social Consciousness

"Transformers" are living proof that the *intangible* exists. Transformers are individuals who intuitively know how to access the higher information sources. In turbulent times unique creative ideas tend to manifest in response to the increase of negative social pressures. During such periods of chaos, transformers work together to form critical masses, so their ideas can surface and be implemented in new ways. In essence, this effect can be viewed as "a noble revolution" which is slowly giving birth to a new awakening for humanity.

When we agree to work in concert with the laws of the universe we become their master, so they work their magic for us. If, however, we disregard these laws we will soon discover that our lives will soon become a mess. Experienced transformers thus tend to work within a *proactive* mindset, rather than in a *reactive* one. This way, they can serve to guide others to achieve their unrealized potentials. Transformers work in the future *now* to form new future scenarios and bring them directly into the present reality. We live in a "monkey-see-monkey-do" world. Thus, if we can be carriers of this dynamic force for change and become skilled at grounding our creative energies into our daily lives, others will seek out our advice and feel more comfortable in supporting our projects and decisions.

Bringing creative energies into the workplace is a challenge that requires constant vigilance, since "resistance to change" permeates the general social consciousness. This is why it is so important to use our intuitive abilities to head off contrived sabotage by others and to defuse developing polarizations before they can develop into serious problems. Once we discover the direct connection between our inner states of beingness and our outer life experiences, we begin to *feel* the creative process operating within us. This feeling is addicting and exhilarating. It is thus a critical factor for empowering the process of spiritual evolution.

Creating Consciousness Fields for Personal, Social and Global Transformation

At this time in history we stand at the threshold of a revolutionary shift in human consciousness. The "real reality show" format offers tantalizing opportunities for change and transformation, but requires corresponding shifts in consciousness. The main keys to survival, success and happiness lie in our ability to insulate our consciousness and *not* to "buy into" the social overlays of fear, greed and control that presently dominate global society. It is thus important to move from the linear-brain thinking modality into a state of recognizing ourselves as the enlightened beings we truly are becoming. *Recognition* is a divine gift we earn when we abandon old restrictive thoughtforms and embrace new visions for the future. By shifting into a more positive mindset we can bypass interference from the linear mind that tries to bring up old patterns of insecurity, abuse, fear and doubt, which were formed during our childhood years and subsequent life-experiences. By making this simple shift we can become clear channels for the information, wisdom and guidance from the higher dimensions. We can then learn to anchor this new information effectively so it can manifest more easily in the physical realm. Once we establish a bridge between the physical and the higher realms, we will find it easier to rise above the negative restrictions of contemporary social patterns and function effectively within a field of joy, confidence and creative accomplishment.

Aligning ourselves with these higher absolute values is what transformation and ascension are all about. We are actually *becoming* our higher selves. This shift is not a mental thing, but can be recognized when we enter that perfect comfort zone of peace, joy and confidence.

Another key to success in working with the quantum field is to visualize this field as a unified information resource which can be accessed at will to receive guidance for making decisions for our highest good. By accessing the quantum field via the heart we establish a strong two-way communications conduit based on *feeling* and *intuition*. This connection will strengthen and refine itself over time as we continue to work with it and learn from constructive feedback. *Intention* and *mental focus* are also important in the process of reinforcing our intuitive antenna systems so we can learn to select those inspirations and guiding coordinates that will most benefit us. Envisioning this two-way cosmic connection thus constitutes a critical tool for establishing and reinforcing these newly opened pathways between our consciousness and the quantum intelligence field. The more we use these new tools, the more easily we can exercise "cosmically intelligent" control over our lives. By accessing the quantum intelligence field we can move beyond our old counterproductive patterns and be empowered to make decisions, which are superior to decisions generated from linear thinking.

Our ego selves have been conditioned by thoughtforms from the past, and also by the restrictive emotional overlays of survival, fright, fight, scarcity and greed. Cruelty and other warlike behavior have repeatedly drained the spiritual and financial resources of entire past civilizations – resulting in their ultimate decline. This has happened over and over again. As a culture, humans identify so powerfully with these restrictive cultural overlays that they are essentially *brainwashed* and unable to see this "great tragedy" which has overshadowed the human race for millennia. By sustaining a healthy heart-mind connection with the cosmic intelligence field we can operate effectively in a zone of "oneness" and with a greater sense of true security. We will be able to experience more peace of mind and freedom from stress. This makes it fun to love and respect ourselves more than ever before, so we can rise above the "human condition" for greater periods of time.

Overcoming Fear and Shifting into a Positive Mindset

From an evolutionary perspective, fear *can* be useful in helping us avoid danger, but when the *rate of change* accelerates fear can become habitual, which causes our emotional bodies to react. Chronic fears often arise in childhood. By the time we become adults these fears tend to remain embedded as part of our social worldview. By acknowledging these fears we become *aware* of the limitations they place on us and can take action to overcome them. This opens the way for gaining wisdom, compassion and a deep appreciation for ourselves. It is also important to realize that our thoughts and emotions function as "energetic patterns" that can influence our health – either positively or negatively. Thus, we cannot separate our thoughts and emotions from our physical well-being, as our thoughts and beliefs about ourselves are important keys for enlightenment and transformation.

Transpersonal experience refers to levels of consciousness, which extend beyond our ego-selves. Transpersonal experiences can vastly expand our sense of personal identity as well as our relationships with humans, plants and the forces of Nature. These experiences also allow us to interact with archetypal, angelic and extraterrestrial beings that can assist us in these shifts if we request it. Transpersonal experiences can be achieved through meditation, but consciousness, resonance and coherence are also major keys to interacting with the quantum field. Eventually, we reach the point where we *no longer* need gurus or religious authorities to tell us *who* we are and *how* to interact with the quantum field.

The shift we are presently experiencing is enabling us to incorporate both the *ordinary* and *extraordinary* into our lives. With the power of wisdom that comes with this shift also comes responsibility – responsibility to ourselves, to others and to the world at large. Transformation is not for the faint of heart, but is truly a "hero's journey." By expanding our consciousness beyond the ego-based limits of awareness we free ourselves to align with higher frequencies. This gives *power* and *meaning* to our individual and collective choices.

Coherence can be consciously empowered and directed. Thus, we can *consciously* sustain coherence with each other and with other life forms. Our mission is not only to repair the damage done by our lack of coherence, but also to evolve ourselves and our world beyond anything ever achieved. Coherence is thus a key strategy for functioning effectively within the quantum field, as it can be *invoked* to effect positive transformations in ourselves and in the world at large. The time is fast approaching when critical masses of humans *will* achieve coherent consciousness. This shift represents a giant leap for mankind, as it will bring us into an entirely new relationship with the Cosmos.

As we begin to work in increasingly expanded states of consciousness, our homework assignment is to learn to integrate our basic brainwave patterns with higher sources. This way, we can sustain a comfortable interaction with others – while continuing to drink from the pools of cosmic wisdom and knowledge. When we shift from linear to quantum thinking modalities we will soon discover we can solve problems instantly and communicate them cogently. We can profoundly affect others through a process I like to call "direct consciousness transfer." This simply involves communication between our own energy fields and the energy fields of others.

Since external and internal coherence are *mutually reinforcing*, a healthy mind promotes a healthy body – and visa versa. According to the sage Sri Aurabindo the next stage in human consciousness will be a "super consciousness." We can function as "catalytic facilitators" for this next state of consciousness by creating bridges between others and ourselves within this super-conscious framework. Thus, we are all learning to become adept at the delicate art of grounding higher knowledge and wisdom into practical applications for the physical world.

Within a super conscious mindset we can better interact with others. Once this new consciousness reaches a critical mass we can move away from a business-as-usual state of mind and break through into new realizations. A critical mass of like-minded individuals constitutes a very powerful force for confronting common dangers and threats to our survival. Is this not what true homeland security is all about? Thus, if we can learn to live eco-sensibly and relinquish our outmoded scarcity, greed and kill-or-be-killed mindsets, this makes it much easier to shift from outmoded linear thinking patterns into a new quantum perspective. In the words of global leader, Ervin Laszlo, "A quantum shift in the global brain is a sudden and fundamental transformation in the relations of a significant segment of the six and half billion humans to each other and to Nature – a Macroshift in society – and a likewise sudden and fundamental transformation in cutting-edge perceptions regarding the nature of reality – a paradigm shift in science. The two shifts together make for a veritable 'reality revolution' in society as well as science" (Laszlo, 2008).

Consciousness Technologies for Ending War and Achieving World Peace

Over the past few decades several different types of consciousness-based defense strategies have been developed. For example, the so-called Spiritual Defense Initiative (SDI) was first put forth by spiritual adept, Swami Satchidananda. In 1978, this led to the formation of the Pentagon Meditation Club, which was funded by the U.S. Department of Defense and involves special "Peacefield Meditations" (Himalayan Academy, 2014). Interestingly, this effort was also quietly supported by a group of Russian leading-edge thinkers, The Russian Initiative Group for Defense of Earth, whose members research the scientific exploration of "Inner Space."

I inadvertently discovered this cooperative effort between these strange bedfellows as follows: In 2003 a Canadian scientific colleague put me in touch with a leading-edge Russian thinker, with whom I collaborated for several years. During this time he translated and summarized many articles on consciousness technology for my own research. In a rather bizarre series of events this Russian researcher provided contact information for American Edward Winchester, whom he had met several years ago at a Buddhist temple in St. Petersburg, Russia. Ed is the original founder of the Pentagon Meditation Club [It really does exist]. Subsequently, I invited Ed to a Conference on Free Energy I was attending in the Washington, DC area. Ed drove up from his home in Virginia, especially to meet us. Over the intervening years he has provided me with several interesting papers, which relate to specific military applications for conflict resolution. More information about Ed Winchester and the Pentagon Meditation Club is available at: www.zoominfo.com/p/Edward-Winchester/195228714

Another conscious-based system that has attracted the attention of various military agencies was called Invincible Defense Technology. IDT is based on the Unified Field Theory, which governs the basic laws of Nature. Since these natural laws are fundamental to the laws that govern the universe, they have the potential to create what has been called an "impenetrable armor of national defense." Since IDT is based on Natural Law it is inherently safe. The basic phenomenon of increased coherence and peace throughout society has been a part of ancient Vedic tradition for thousands of years. It was first brought back to the public attention in 1976.

According to military experts, traditional defense strategies fail because they do not address the underlying causes of violence and conflict. In other words, they do not relieve the acute political, ethnic and religious tensions that fuel this kind of destructive behavior. Instead, what is needed is an effective means to lessen or eliminate these deep-seated tensions – a proven approach for safeguarding a country against violence, which will serve to promote peace and prosperity in the world.

Unknown to most of us, IDT has been extensively field-tested in the Middle East and other parts of the world. Consistent results yielded dramatic reductions in terrorism, war and other social violence. Research trials have been replicated, published in leading peer-reviewed scientific journals and endorsed by hundreds of independent scientists and scholars. Thus, it would appear that the validation of invincible defense techniques is now beyond question. Within this context IDT represents a scientifically proven, practical approach to achieving national stability and world peace. In summary, Invincible Defense Technology constitutes an applied technical application of the latest discoveries in the fields of neuroscience, quantum mechanics and consciousness technology (Hagelin and Leffler, 2010).

This new "peace technology" focuses on establishing large national groups of trained individuals, specifically trained in the IDT techniques which have already proven effective in neutralizing ethnic, political and religious tensions that have historically generated conflict and terrorism. Since IDT works at the deepest and most powerful levels of Natural Law, it has the power to effectively render ineffective conventional military technologies such as chemical, biological, electronic and nuclear weapons. This was validated when researchers discovered that when 1% of a community practiced the Transcendental Meditation program, crime rates dropped by an average of 16%. This phenomenon was named the "Maharishi Effect" in honor of Maharishi Mahesh Yogi, who first predicted the effect in 1960. The term Maharishi Effect now most often refers to the influence generated by group practice of the advanced TM-Sidhi program, through which *even the square root of 1% of a community* is sufficient to create measurable changes in social trends.

In 2007, quantum physicist and public policy expert, Dr. John Hagelin, inaugurated the International Center for Invincible Defense Technology in New York. This center serves as international headquarters for a new global initiative to implement IDT. The center's primary objective is to establish permanent peace-creating groups to practice advanced peace-meditation techniques in every nation of the world. Hagelin claims that the practice of IDT by groups of peace-creating experts has the power to activate The Unified Field – the most fundamental and powerful organizational level of Nature. He also claims that the positive, coherent influence created by these groups can calm international hostilities to prevent war and terrorism. Dr. Hagelin, states: "During the last 25 years, Invincible Defense Technology has been extensively field tested in the Middle East and throughout the world." He goes on to say, "The consistent result has been dramatic reductions in terrorism, war, and other social violence. These findings have been replicated, published in leading academic journals and endorsed by hundreds of independent scientists and scholars. The efficacy of this approach is now beyond question" (International Center for Invincible Defense Technology, 2013).

In conclusion, the main effects of IDT in practice include: 1) Defusing enmity in potential adversaries, which serves to prevent the creation of enemies, conflict escalation and the eventual onset of war; 2) Creating a coherent and harmonious national consciousness, which effectively strengthens a nation by eliminating disruptive influences both externally and internally; 3) Creating a peaceful and cooperative global environment, which encourages the formation of stable governments and cessation of hostilities; and 4) Establishing lasting peace, security and economic stability at home. This would relieve the tremendous pressures on government institutions charged with public safety and homeland security. Invincible Defense Technology is claimed to be the only approach to peace and conflict resolution that effectively targets the root causes of regional, national and global conflict. IDT is also said to be the only approach to national security and world peace that is extensively field tested and backed by scientific research. In terms of military defense this unified field-based technology has already proven to effectively in generating a measurable "peace effect" which creates a positive environment within which diplomatic and other conflict resolution approaches can actually succeed. For these reasons, it is important that this new peace technology should be the basic component of any comprehensive approach to peace and national security. A list of links to national military groups and other organizations using IDT to promote peace to can be found at: www.davidleffler.com

Finally, it is important to point out that one of the more appealing aspects of Invincible Defense Technology is that non-military groups now have the ability to initiate positive social change through group meditations and musical performance events. Ideally, world religions could integrate world peace meditations into their traditional worship services, as could businesses, schools and government offices.

Closing Message from the Author

The 25 Master Keys presented in the foregoing pages are based on years of extensive experience and research in many different fields. It was my intent to present these master paradigms as "seeds" for others to germinate and harvest in their own quest to improve various aspects of the "human condition." It is hoped that this harvest can be shared and refined by future generations. These master keys are all related so that each key is an integral part of the tapestry of Future-Science Technology.

It is important to remember, what is perceived as a miracle is most often the result of countless tiny steps in the right direction. We are making miracles every day. In the Disney classic movie, *Pinocchio*, Jiminy Cricket sings a little song called, "Let Your Conscience Be Your Guide." When you allow intuition to be your guide, you are well on your way to working with the quantum field. Never be afraid to push your limits to the next level. Never, ever give up, as to do so would dishonor those who suffered and died to pave our way. Entertain the outrageous, celebrate your success and honor the achievements of others. Share your happiness by offering your knowledge to others. Surround yourselves with friends and colleagues of like mind. When conflict arises in your life, learn to understand the worldviews of others and to "take a walk in the other person's moccasins." If you decide to shoot the messenger, understand that you will get no more mail. May your creative forge be stoked by the fires of the gods! Enjoy your adventures in the Quantum Fields and celebrate your own Brave New Mind!

> "…Beyond the 'tomorrowland' destined to blossom upon the earth plane
> lies the timeless realm of oneness with the God Force,
> to which you have been extended a tentative invitation.
> The RSVP is a never-ending decision,
> made and remade with every nuance of every waking moment.
> It is an invitation you carry with you at all times engraved in the structure of your genetic code, and printed out beautifully in every cell of your physical being.
> Blessed be he who recognizes and honors self as a boundless expression of that knowningness, and never questions that he is eminently worthy of the distinction…
> with humblest gratitude that this is so."
>
> *Rasha, The Calling*

About the Author

Dr. Elliott Maynard is a neo-renaissance, leading-edge consciousness scientist and conceptual designer who walks his talk, lives his dream and is driven to share this information with the world. His background spans the fields of Global Ecology, Coral Reef Ecology, Oceanography and Tropical Rainforest Biology. He earned a Ph.D. in Consciousness Research and has served on the faculties of Adelphi University and Dowling College in New York. He is a Certified Professional Consultant to Management (CPCM) and has been active in the corporate world as Founder, President and Technology Director for several different corporations. Dr. Maynard is Founder and President of Arcos Cielos Foundation in Sedona, Arizona and has been active in the Aerospace Technology Working Group (ATWG), the Humanitad International Leadership Foundation, The U.S. Psychotronics Association and the World Future Society. He also serves on the editorial board for the Kepler Institute for Space Philosophy.

In addition to his work as a futurist, lecturer, author, educator and global ecologist, Elliott is an accomplished artist, sculptor, musician, underwater photographer and documentary filmmaker. Because of his unique abilities to bring future concepts into the present and integrate them into the existing social structure, he has been referred to by some of his colleagues as "The Human from the Future."

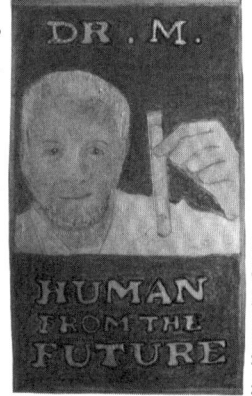

BIBLIOGRAPHY

In addition to the references cited in the text, I have also provided a number of additional references, which have formed the foundations of Future-Science Technology over the past years. These books have been carefully selected for their excellence and general reading enjoyment. The best way to utilize this is to simply browse through the following list of references and highlight the topics that interest you. These references should also be useful for students and researchers wishing to learn more about specific aspects of the included sections and to update and expand the basic concepts.

Aburdene, Patricia and J. Naisbitt, 1992 – *Megatrends for Women*. Villard Books, NY.

Abrams, Lindsay, 2013 – "Does Russ George Deserve a Nobel Prize or a Prison Sentence?" Salon, August 30, 2013, www.salon.com/2013/08/30/does_millionaire_russ_george_deserve_a_nobel_prize_or_a_ prison_sentence

Adair, David, 2012. *Area 51: Advanced Symbiotic Technology*. www.youtube.com/watch?v=M0wBXNAiOys&NR=1&feature=fvwp

Adachi, Ken, 2003 – "Trevor James Constable, A Man of Seasons." Educate-Yourself, January 6, 2003, www.educate-yourself.org/tic/briefbio.shtml

Adams, John, 1974 – *Conceptual Blockbusting: A Guide to Better Ideas*. W. W. Norton & Company, NY.

Adams, John, 2000 – *Thinking Today as if Tomorrow Mattered: The Rise of a Sustainable Consciousness*. Eartheart Enterprises, San Francisco, CA.

Advanced Magnetic Research Institute, 2014 – "Magnetic Molecular Energizer (MME)." www.amri-intl.com

Agor, Weston H., 1989 – *Intuition in Organizations: Leading and Managing Productivity*. Sage Publications, London.

Alexander, Eben, 2012 – *Proof of Heaven: A Neurosurgeon's Journey into the Afterlife*. Simon and Schuster, NY.

Allen, Pat B., 1995 – *Art as a Way of Knowing: A Guide to Self-Knowledge and Spiritual Fulfillment through Creativity*. Shambhala Publications, Inc., Boston, MA.

Allen, Phil, A. Bearne and R. Smith, 1977 – *Energy, Matter and Form: Toward a Science of Consciousness*. University of the Trees Press, Boulder Creek, CA.

Amos, Jonathan, 2012 – "World Glaciers 'Out of Balance.'" BBC News: Science & Environment, April 26, 2012.

Anderson, George and A. Barone, 1999 – *Lessons from the Light: Extraordinary Messages of Comfort and Hope from the Other Side*. Berkeley Publishing Group, NY.

Andrews, Ted, 1994 – *How to Meet and Work with Spirit Guides*. Llewllyn Publications, Woodbury, MN.

Antoniades, Andri, 2013 – "Can We Replace Lost Coral Reefs with 3D Printed Versions?" November 7, 2013.

Aronson, Virginia, 1999 – *Celestial Healing: Close Encounters That Cure*. Penguin Putnam, Inc., NY.

Atwater, P. M. H., 1996 – "The Future-Memory Phenomenon." www.cinemind.com/atwater/futmem.html

Atwater, P.M.H., 1999 – *Future Memory: How Those Who "See the Future" Shed New Light on the Working of the Human Mind*. Birch Lane Press, NY.

Atwater, P.M.H., 1999 – *Children's Near-Death Experiences and the Evolution of Humankind*. Three Rivers Press, NY.

Atwater, P. M. H., 2004 – *We Live Forever: The Real Truth About Death, With Wisdom From the Edgar Casey Readings*. A.R.E. Press, Virginia Beach, VA.

Atwater, P. M. H. 2007 – *The Big Book of Near-Death Experience: The Ultimate Guide to What Happens When We Die*. Hampton Roads Publishing Company, Inc., Charlottesville, VA.

Baldwin, William J., 2009 – *Spirit Releasement Therapy: A Technique Manual*. Headline Books, Terra, Alta, WVA.

Barnaby, Frank, 1988 – *The Gaia Peace Atlas: Survival into the Third Millennium*. Doubleday & Co, Inc., NY.

Bashar, 2012 – "Entity channeled from the higher dimensions by Darryl Anka." www.bashar.org

BBC News, 2012b – "Will the Internet Become Conscious?" November 22, 2012. www.bbc.com/future/story/20121121-will-the-net-become-conscious

BBC News, 2013 – "3D Printed Moon Building Designs Revealed." BBC News: Technology, February 1, 2013. www.bbc.co.uk/news/technology-21293258

Beachley, 2012 – "A Current Affairs Follow-Up Report on SCENAR Therapy." YouTube Video, www.youtube.com/watch?v=qQ-dGCQvXxM

Bearden, Thomas E., 1988 – *The Excalibur Briefing: Explaining the Paranormal Phenomenon*. Strawberry Hill Press, San Francisco, CA.

Bearden, Thomas E., 2002 – *Energy from the Vacuum: Concepts and Principles*. Cheniere Press, Santa Barbara, CA.

Bearden, Thomas E., 2009 – "Precursor Engineering." www.zpenergy.com/modules.php?name=News&file=article&sid=3064

Bearden, Thomas E., 2005 – *Oblivion: America at the Brink*. Cheniere Press, Santa Barbara, CA.

Beaulieu, John, 1987 – *Music and Sound in the Healing Arts: An Energy Approach*. Station Hill Press, Inc., Barrytown, NY.

Before It's News, 2011 – "HAARP Facilities Worldwide." Before It's News, Jan. 12, 2011. www.beforeitsnews.com/alternative/2011/01/haarp-facilities-worldwide-353828.html

Begich, Nick and J. Manning, 1995 – *Angels Don't Play This Harp: Advances in Tesla Technology*. Earthpulse Press, Anchorage, AK.

Begich, Nick, 2006 – (DVD). *HAARP – The Update: Angels Still Don't Play this HAARP*. www.earthpulse.com

Bennett, Bonnie C., 2014 – Bonnie Bennett, D. O., M. P. H. www.easystreetcoaching.com/v-i-p-profiles/bonnie-c-bennett

Bentov, Itzhak, 1977 – *Stalking the Wild Pendulum: On the Mechanics of Consciousness*. Destiny Books, Rochester, VT.

Bentov, Itzhak, 1988 – *A Cosmic Book: On the Mechanics of Creation*. Destiny Books, Rochester, VT.

Bethards, Betty and J. Grace, 1998 – *Seven Steps to Developing Your Intuitive Powers*. Element Books, Inc., Rockport, MA.

Biello, David, 2012a – "A Rogue Climate Experiment Outrages Scientist." New York Times, October 18, 2012.

Biello, David, 2012b – "Can Controversial Ocean Iron Fertilization Save Salmon?" *Scientific American*, October 24, 2012. www.scientificamerican.com/article/fertilizing-ocean-with-iron-to-save-salmon-and-earn-money

Bird, Christopher, 1993 – *The Divining Hand: The 500-Year-Old History of Dowsing*. Shiffer Publishing, Ltd., Atglen, PA

Bissler, Jane V., D. Florino and S. Ruble, 2008 – *Surviving and Thriving: Grief Relief and Continuing Relationships*. www.spiritualityworkshops.com

Bissler, Jane V. and L. Heiser, 2008 – *Loving Connections: The Healing Power of After-Death Communications*. Counseling for Wellness, LLP, 420 W. Main St., Kent, OH.

Blackburn, Gabriele, 1983 – *The Science and Art of Pendulum: A Complete Course in Radiesthesia*. Idylwild Books, Ojai, CA.

Bloom, Allan, 1987 – *The Changing of the American Mind*. Simon & Schuster, NY.

Bonlie, Dean R., 2012 – "Bio-Magnetic Theory." www.shokos.com/science.htm

Boschmann, Erwin, Ed., 1995 – *The Electronic Classroom: A Handbook for Education in the Electronic Environment*. Learned Information, Inc., Medford, NJ.

Botkin, Allan L., 2005 – *Induced After Death Communication: A New Therapy for Healing Grief and Trauma*. Hampton Roads Publishing Company, Inc., Charlottesville, VA.

Boulding, Kenneth E., 1985 – *The World as a Total System*. Sage Publications, Beverley Hills, CA.

Braden, Greg, 1997 – *Awakening to Zero Point: The Collective Initiation*. Radio Bookstore Press, Belleview, WA.

Braden, Greg, 2007 – *The Divine Matrix: Bridging Time, Space, Miracles, and Belief*. Hay House, Inc., Carlsbad, CA.

BrainWell Center, 2013 - www.brainwellcenter.com

Brand, Stewart, 1987 – *The Media Lab: Inventing the Future at MIT*. Penguin Books, NY.

Brown, Thomas E., 1994 – *The Loom of the Future: The Weather Engineering Work of Trevor James Constable*. Borderlands Sciences Research Foundation, Eureka, CA. www.borderlands.com/constable.htm

Burks, Fred, 2013 – "HAARP: Secret Weapon Used for Weather Modification, Electronic Warfare." Global Research, July 23, 2013.

Buzon, Tony, 1974 – *Use Both Sides of Your Brain: New Techniques to Help You Read Efficiently, Study Efficiently, Solve Problems, Remember More, Think Creatively*. E. P. Dutton, NY.

Callahan, Philip S., 1975 – *Tuning Into Nature: Solar Energy, Infrared Radiation and the Insect Communication System*. The Devin-Adair Company, Old Greenwich, CT.

Callahan, Philip S., 1994 – *Exploring the Spectrum: Wavelengths of Agriculture and Life*. Acres USA, Kansas City, MO.

Callahan, Philip S., 1995 – *Paramagnetism: Rediscovering Nature's Secret Force*. Acres USA, Metairie, LA.

Cameron, Julia, 1996 – *The Vein of Gold: A Journey to Your Creative Heart*. Jeremy P. Tarcher/Putnam, NY.

Cameron, Julia, 1996 – *The Zen of Creative Painting*. Watson-Guptill Publications, NY.

Campbell, Robert, 1985 – *The Fisherman's Guide: A Systems Approach to Creativity*. Shambhala Publications, Boston, MA.

Capra, Fritjof, 1976 – *The Tao of Physics*. Bantam Books, NY.

Capra, Fritjof, 1982 – *The Turning Point: Science, Society, and the Rising Culture*. Bantam Books, NY.

Carlsberg, Kim, 2010 – *The Art of Close Encounters*. Close Encounters Publishing, www.closeencounterspublishing.com

Cassou, Michele and S. Cubley, 1995 – *Life, Paint and Passion: Reclaiming the Magic of Spontaneous Expression*. Jeremy P. Tarcher/Putnam, NY.

Cathie, Bruce L., 1998 – *The Harmonic Conquest of Space*. Adventures Unlimited Press, Kempton, IL.

CBC News, 2008 – "HAARP CBC Broadcast Weather Control." Sept. 18, 2008, www.youtube.com/watch?v=10DihgRIs0I&list=PL99AB5E3E70406BF0

CBS News, 2012 – "Modern Wheat: A Perfect Chronic Poison, Doctor Says." www.cbsnews.com/8301-505269_162-57505149/modern-wheat-a-perfect-chronic-poison-doctor-says

Center for Spiritual Understanding, 2013 – "Self-Guided Afterlife Connections." www.spiritualunderstanding.org

Center for the Study of Extraterrestrial Intelligence, P. O. Box 265, Crozet, VA 22932. www.CSETI.org

Cetron, Marvin and O. Davies, 1989 – *American Renaissance: Our Life at the Turn of the 21st Century*. St. Martin's Press, NY.

Cetron, Marvin and O. Davies, 1991 – *The Crystal Globe: The Haves and Have-Nots of the New World Order*. St. Martin's Press, NY.

Cetron, Marvin and M. Gayle, 1991 – *Educational Renaissance: Our Schools at the Turn of the Twenty-First Century*. St. Martins Press, NY.

Chernoff, Allan, 2010 – "Enviropig: The Next Transgenetic Food." www.eatocracy.cnn.com/2010/09/25/enviropig-the-next-transgenic-food

Choate, Trish, 2011 – "Scientists Studying Wind Farms Effects on Temperature, Weather." *Standard Times,* June 17, 2011.

Chopra, Deepak, 1993 – *Ageless Body, Timeless Wisdom: The Quantum Alternative to Growing Old*. Harmony Books, NY.

Christensen, Sallie with G. M. Hillier, 2000 –*The Highest and the Best: A Gifted Healer's Vision of Third-Millennium Medicine and Humanity's Intuitive Evolution*. Self-Published by the authors, P.O. Box 8, Harmony, PA 16037.

Church, George M. and E. Regis, 2012 – "Repelling Viruses, Reviving Mammoths." *Wall Street Journal, October 19, 2012.* www.online.wsj.com/news/articles/SB10000872396390443684104578062734031207640

Cohen, Frederick B., 1994 – *It's Alive: The New Breed of Computer Programs*. John Wyley & Sons, NY.

Coldewey, Devin, 2012 – "Toddler Calls 3D Printed Medical Exoskeleton 'Her Magic Arms.'" NBC News, www.nbcnews.com/technology/futureoftech/toddler-calls-3-d-printed-medical-exoskeleton-her-magic-arms-928612

Collinge, William, 1998 – *Subtle Energy: Awaking to the Unseen Forces in Our Lives*. Warner Books, Inc., NY.

Connor, Caroline, 1994 – *Instant Radionics: Educational Radionic Workbook and General Information*. International Guild of Advanced Sciences College Research Society, Palm Springs, CA.

Constable, Trevor James, 2008 - *The Cosmic Pulse of Life: The Revolutionary Biological Power behind UFOs*. Borderlands Sciences Research Foundation, Eureka, CA. www.borderlands.com/home.htm

Cramer, Guy, 2007 – "HAARP Transmissions May Accidentally Be Jamming Bee's Homing Ability." Superstealth.com, June 1, 2007. www.hyperstealth.com/haarp/index.htm

Currivan, Jude, 2006 – *The 8th Chakra: What It Is and How It Can Transform Your Life*. Hay House, Inc., Carlsbad, CA.

Da Free John, 1978 – *The Enlightenment of the Whole Body*. Dawn Horse Press, Middletown, CA.

Dalrymple, Ron, 1987 – *The Inner Manager: Mastering Business, Home, and Self*. Celestial Gifts Publishing, Chester, MD.

Dateline NBC, 1996 – "Dateline NBC April 19, 1996 Transcript of Edgar Mitchell Interview." www.v-j-enterprises.com/dvmtchl1.html

Davidson, John, 1987 – *Subtle Energy*. C. W. Daniel Company, Ltd., London, UK.

Davidson, John, 1987 – *The Web of Life Force: The Energetic Constitution of Man and the Neuro-Endocrine Connection*. C. W. Daniel Co, Ltd., UK.

Dean, Robert O., 2013 - Command Sergeant-Major, USAF (Ret.); Intelligence Agent (Cosmic Top Secret Clearance), Supreme Headquarters, Allied Powers, Europe. www.fastwalkers.com/featured/RobertDean.htm

Denning, Melita and O. Phillips, 1988 – *The Llewellyn Practical Guide to Psychic Self-Defense and Well-Being*. Llewellyn Publications, St. Paul, MN.

Diep, Francie, 2010 – "How 3D Printing Could Become Commonplace." *Innovation-News Daily*, Aug. 8, 2012, www.technewsdaily.com/6084-3d-printing-commonplace.html

Disclosure Project, The – P. O. Box 265, Crozet, VA 22932, www.DisclosureProject.org

Divine Life Foundation, 2010 – www.divinelifefoundation.us/Transcendental_Science/Transcendental_Science.htm

Dizard, Wilson, 1997 – *Meganet: How the Global Communications Network Will Connect Everyone on Earth*. Westview Press, Boulder, CO.

Dolnick, Barrie, 1998 – *The Executive Mystic: Psychic Power Tools for Success*. HarperCollins Publishers, NY.

Dong, Paul T., E. Raffill and K. Kramer, 1997 – *China's Super Psychics*. Marlowe and Company, NY.

Dossey, Larry, 2002 – "Measuring the Power of Prayer." *Whole Earth*, Spring, 2002.

Duchamp, Mark, 2012 – "Wind Turbine Syndrome Affects More People Than Previously Thought." WUWT (Watts Up With That?), March 6, 2012: www. wattsupwiththat.com/2012/03/06/wind-turbine-syndrome-affects-more-people-than-previously-thought

Durrell, Lee, 1986 – *State of the Ark: An Atlas of Conservation in Action*. Doubleday & Company, NY.

Dvir, Adrian, 2003 – *X3, Healing, Entities, and Aliens*. Book Masters, Inc., Rishon Lezion, Israel. www.adriandvir.com

Dyson, Freeman, 1997 – *Imagined Worlds*. Harvard University Press, Cambridge, MA.

Eason, Cassandra, 2005 – *The Psychic Power of Children: How to Deal With It*. Foulsham & Company, Ltd., NY.

Edwards, Alexis, 1971 – *Guidelines*. Findhorn Press, Inc., Scotland.

Edwards, Betty, 1979 – *Drawing With the Right Side of Your Brain*. J. P. Tarcher, Inc., Los Angeles, CA..

Eilperin, Juliet, 2012 – "Greenland Ice Sheet Had Biggest Thaw Since 2006" – (DVD).

Escamilla, Jose, 2006 – *UFO: The Greatest Story Ever Denied*. www.theUFOmovie.com

Essene, Virginia and S. Nidle, 1994 – *You Are Becoming a Galactic Human*. Spiritual Endeavors Publishing Company, Santa Clara, CA.

Exopolitics, 2013 – Exopolitics.com, www.exopolitics.com

Fabin, Don, 1971 – *Telecommunications: One World Mind*. Glencoe Press, Beverly Hills, CA.

Farsight Institute, The, 2013 – Institute for Remote Viewing. www.farsight.org/index.html

Ferguson, Marylyn, 1980 – *The Aquarian Conspiracy: Personal and Social Transformation in the 1980's*. J. P. Tarcher, Inc., Los Angeles, CA.

Fisk, Edward B., 1992 – *Smart Schools, Smart Kid: Why Do Some Schools Work?* Simon and Schuster, NY.

Fountain, Henry, 2012 – "A Rogue Climate Scientist Outrages Scientist." *New York Times*, October 18, 2012.

Fox News, 2008 – "Ex-Astronaut: Aliens Are Real and Government Knows It." Newscorp Australian Papers, July 25, 2008," www.foxnews.com/story/2008/07/25/ex-astronaut-aliens-are-real-and-government-knows-it/#ixzz2ZzX2bNru

Futurist, 2012 – "Market for Bioplastics: Businesses Are Developing Green Alternatives to Fossil-Fuel Plastics." *Futurist*, November-December, 2012.

Gallagher, James, 2014 – "Stem Cell 'Major Discovery' Claimed." BBC News, January 29, 2014. www.aggbot.com/Headline-News/article/21889181

Gallwey, Tim and R. Kriegel, 1997 - *Inner Skiing*. Random House, NY.

Gelernter, David, 1991 – *Mirror Worlds: The Day Software Puts the Universe in a Shoebox…How It Will Happen and What It Will Mean*. Oxford University Press, NY.

Gismodo, 2013 – "Engineering 3D Printed Stem Cells," March 22, 2013, www.gizmodo.co.uk/2013/03/engineering-3d-printed-stem-cells

Glatzer, Hal, 1983 - *The Birds of Babel: Satellites and Their Impact on the Environment*. Howard W. Sams & Co., Inc., Indianapolis, IN.

Godlike Productions, 2014 – "UFO's Are Real: Ben Rich Lockheed Skunk Works Director Admitted in His Deathbed Confession." April 1, 2014. www.godlikeproductions.com/forum1/message2079877/pg1

Gold, Aviva, 1998 – *Painting from the Source: Awakening the Artist's Soul in Everyone*. HarperCollins Publishers, Inc., NY.

Goldberg, Bruce, 1997 – *Peaceful Transition: The Art of Conscious Dying & the Liberation of the Soul*. Llewellyn Publications, St. Paul, MN.

Goldberg, Philip, 1983 – *The Intuitive Edge: Understanding Intuition and Applying It to Everyday Life*. Jeremy P. Tarcher, Inc., Los Angeles, CA.

Goodman, T, 2010 – "Whale Poop is Vital to Ocean Ecology." www.inventorspot.com/articles/whale_poop_vital_ocean_ecology

Goodspeed, Bennett, 1983 – *The Tao Jones Averages: A Guide to Whole-Brained Investing*. E. P. Dutton, Inc., NY.

Gotz, Frankie, 2013 – "Major Flooding in Alberta – Could It Be Linked to Geoengineering?" Geoengineering Watch, June 28, 2013. www.geoengineeringwatch.org/major-flooding-in-alberta-could-it-be-linked-to-geoengineering

Graci, Sam, 1999 – *The Power of Superfoods*. Prentice Hall Canada, Inc., Scarborough, Ontario.

Gray, Louise, 2012 – "Wind Farms Can Cause Climate Change, Finds New Study." *The Telegraph*, April 29, 2012. www.telegraph.co.uk/earth/earthnews/9234715/Wind-farms-can-cause-climate-change-finds-new-study.html

Greenberg, Paul, 2011 – *Four Fish: The Future of the Last Wild Food*. Penguin Books, NY.

Greer, Steven M., 1999 – *Extraterrestrial Contact: The Evidence and Implications*. Granite Publishing, LLC, Columbus, NC.

Greer, Steven M., 2001 – *Disclosure: Military and Government Witnesses Reveal the Greatest Secrets in Modern History*. Crossing Point, Inc., Crozet, VA.

Greer, Steven M., 2009 – *Contact: Countdown to Transformation: The CSETI Experience 1992-2009*. 123 PrintFinder, Inc., Ladera Ranch, CA 92694.

Grey, Alex, 1998 – *The Mission of Art*. Shambhala Publications, Inc., Boston, MA.

Griffin, Edward, M. Murphy and P. Wittenberger, 2010 - *What In the World Are They Spraying? The Chemtrail/Geoengineering Cover-Up*. (DVD). Reality Zone, www.realityzone.com

Guggenheim, Bill and J. Guggenheim, 1995 – *Hello From Heave: A New Field of Communication Confirms That Love and Life Are Eternal*. Bantam Books, NY.

HAARP.net, 2024 – Hagelin, John and D. Leffler, 2010, www.globalgoodnews.com/world-peace-a.html?art=126280444818949208

Hall, Judy, 1997 – *The Art of Psychic Protection*. Samuel Weiser, Inc., York Beach, ME.

Hardcastle, Rebecca, 2008 – *Exoconsciousness: Your 21st Century Mind*. Author House, Bloomington, IN.

Harder, Heather Anne, 1993 – *Exploring Life's Last Frontier: The World of Death, Dying and Letting Go*. Channel One Communications, Inc., Needham, MA.

Harmon, Willis, 1988 – *Global Mind Change: The Promise of the Last Years of the Twentieth Century*. Knowledge Systems, Inc., Indianapolis, IN.

Harmon, Willis, and H. Rheingold, 1984 –*Higher Creativity: Liberating the Unconscious for Breakthrough Insights*. Jeremy P. Tarcher, Los Angeles, CA.

Harrabin, Roger, 2012 – "Liquid Air Offers Energy Storage Hope." BBC News, October 1, 2012, www.bbc.co.uk/news/science-environment-19785689

Hastings, Robert, 2008 – *UFOs and Nukes: Extraordinary Encounters at Nuclear Weapons Sites*. AuthorHouse, Bloomington, IN.

Hartman, Thom, 1998 – *The Last Hours of Ancient Sunlight: Waking up to Personal and Global Transformation*. Mythical Books, Northfield, VT.

Henderson, Hazel, 1996 – *Building a Win-Win World: Life Beyond Global Economic Warfare*. Berrett-Koehler Publishers, San Francisco, CA.

Hendricks, Gay and K. Ludeman, 1996 – *The Corporate Mystic: A Guidebook for Visionaries with Their Feet on the Ground*. Bantam Books, NY.

Herzing, Denise, 2004 – "Aquatic Culture – Dolphins Communication." First Science.com, www.firstscience.com/home/articles/nature/aquatic-culture-dolphins-communication_1358.html

Herbert, Nick, 1985 – *Quantum Reality: Beyond the New Physics: An Excursion into Metaphysics*. Anchor/Doubleday, NY.

Hills, Christopher, 1977 – *Nuclear Evolution: Discovery of the Rainbow Body*. University of the Trees Press, Boulder Creek, CA.

Hills, Christopher, 1975 – *Supersensonics: The Supersensitive Life of Man*. University of the Trees Press, Boulder Creek, CA.

Hills, Christopher, 1979 – *The Golden Egg: Manifesting the Rise of the Phoenix*. University of the Trees Press, Boulder Creek, CA.

Hills, Christopher, 1979 – *The Rise of the Phoenix: Universal Government by Nature's Laws*. University of the Trees Press, Boulder Creek, CA.

Hirsch, E. Ed. Jr., 1996 – *The Schools We Need: Why We Don't Have Them*. Doubleday, NY.

History Channel, 2013 – *Weather Warfare Documentary* (Video). February 11, 2013. www.youtube.com/watch?v=WgzFWy2K_M8

Hopewell, Luke, 2012 – "3D Printers Build You a House in 20 Hours: Welcome to the Future." *Gizmodo*, 8/9/12, www.gizmodo.com.au/2012/08/3d-printer-can-build-you-a-house-in-20-hours-welcome-to-the-future

Horowitz, Leonard, G and J. S. Puleo, 1999 – *Healing Codes for the Biological Apocalypse*. Tetrahedron Publishing, LLC, Las Vegas, NV.

Horowitz, Leonard G., 2004 – *DNA: Pirates of the Sacred Spiral*. Tetrahedron Publishing Group, Sandpoint, ID.

Houston, Jean, 1982 – *The Possible Human: A Course in Enhancing Your Physical, Mental, and Creative Abilities*. J. P. Tarcher, Inc., Los Angeles, CA.

Houston, Jean, 2000 – *Jump Time: Shaping Your Future in a World of Radical Change*. Jeremy P. Tarcher/Putnam, NY.

Hubbard, Barbara M., 1998 – *Conscious Evolution: Awakening the Power of Our Social Potential*. New World Library, Novato, CA.

Huffington Post, 2013 – "3D Printing: No Longer Just an Idea." December 7, 2013. www.huffingtonpost.com/jackson-mariotti/3d-printing-no-longer-jus_b_4400731.html

Hughes, Virginia, 2013 – "Will This Fish Transform Medicine? Why the Tiny Zebrafish is Becoming Researchers' Favorite Animal." *Popular Science*, February, 2013.

Hulse, David, 2012 – "Forgotten in Time: The Ancient Solfeggio Frequencies." www.somaenergetics.com/forgotten_in_time.php

Hunt, Valerie V., 1989 – *Infinite Mind: The Science of Human Vibrations*. Malibu Publishing Co., Malibu, CA.

Hutchinson, Michael, 1996 - *Mega Brain Power: Transform Your Life With Mind Machines and Brain Nutrients*. Ballantine Books, NY.

International Center for Invincible Defense, 2013 - www.invincibledefense.org

Ivanenko, Constantin I., 2003 – Founder of Russian Initiative for Defense of Earth, St. Petersburg, Russia, Personal Communication, 2001-2003.

James, Ron and J. Stein, 2011 – (DVD). *The Disclosure Dialogues: A Historical Achievement on Five DVDs*. www.SedonaMediaCompany.com

Jean, Grace V., 2010 – "Converting Eco-Friendly Plastic into Fuel." *National Defense Business and Technology Magazine*, June, 2010.

Jones, Glenn R., 1996 – *Cyberschools: An Education Renaissance*. Jones Digital Century, Inc., Englewood, CO.

Jonscher, Charles, 1999 – *The Evolution of Wired Life: From the Alphabet to the Soul-Catcher Chip – How Information Technologies Change Our World*. John Wiley & Sons, NY.

Kanter, Rosabeth M., 1983 – *The Change Master: Innovation & Entrepreneurship in the American Corporation*. Simon & Schuster, NY.

Kappraff, Jay, 1991 – *Connections: The Geometric Bridge between Art and Science*. McGraw-Hill, Inc., NY.

Kean, Leslie, 2010 – *UFOs: Generals, Pilots, and Government Officials Go on the Record*. Harmony Books, NY.

Keane, Phillip, 2013 – "Another Asteroid Mining Company Opens Shop." *Space Safety Magazine*, January 23, 2013. www.spacesafetymagazine.com/2013/01/23/asteroid-mining-company-opens-shop

Kearney, Brendan, 2012 – "Ridgeville Biotech Firm ArborGen Buys Assets of Rival Tree Seedling Supplier." *The Post and Courier,* August 15, 2012. www.postandcourier.com/apps/pbcs.dll/article?AID=/20120815/PC05/120819528

Keith, David, 2010 – "Subcommittee on Energy and Environment Hearing – Geoengineering II: The Scientific Basis and Engineering Challenges." www.science.house.gov/hearing/subcommittee-energy-and-environment-hearing-geoengineering

Kelly, Penny, 1997 – *The Elves of Lilly Hill Farm: A Partnership with Nature*. Llewellyn Publications, St. Paul, MN.

Kerrang Radio 2008 - Ed Mitchell YouTube Radio Interview. www.youtube.com/watch?v=RhNdxdveK7c

Keyes, Ken, 1982 – *The Hundredth Monkey*. Vision Books, Coos Bay, OR.

King, Serge Khalil, 1992 – *Earth Energies: A Quest for the Hidden Powers of the Planet*. Quest Books, Wheaton, IL.

Kiyosaki, Robert T. and S. L. Lechter, 1997 – *Rich Dad, Poor Dad: What the Rich Teach Their Kids About Money – That the Poor and Middle Class Do Not!* Warner Books, NY.

Klimo, Jon, 1998 – *Channeling: Investigations on Receiving Information from Paranormal Sources*. North Atlantic Books, Berkeley, CA.

Koepf, Herbert H., 1989 – *The Biodynamic Farm*. Anthrosophic Press, Hudson, NY.

Krech, Shepard, 1999 – *The Ecological Indian: Myth and History*. W. W. Norton & Company, NY.

Krivit, Steven B., 2007 – "The Mistake of Pons and Fleischmann and Why Their Discovery Was Initially Thought to Be a Mistake." *New Energy Times*, March 23, 2007. www.newenergytimes.com/v2/reports/MistakesOfFleischmannAndPons.shtml

Kyron, 2010 – *The Twelve Layers of DNA: An Esoteric Study of the Mystery Within*. Platinum Publishing House, Sedona, AZ.

Kyron, 2012 – Interdimensional entity channeled by Lee Carroll. www.kryon.com

Laszlo, Ervin, 2004 – *Science and the Akashic Field: An Integral Theory of Everything*. Inner Traditions, Rochester, VT.

Laszlo, Ervin, 2008 – *The Quantum Shift in the Global Brain: How the New Scientific Reality Can Change Us and Our World*. Inner Traditions, Rochester, VT.

Laszlo, Erwin, 2009 – *WorldShift 2012: Making Green Business, New Politics & Higher Consciousness Work Together*. Inner Traditions, Rochester, VT.

Laszlo, Erwin, 2009 – *The Akashic Experience: Science and the Cosmic Memory Field*. Inner Traditions, Rochester, VT.

Laszlo, Ervin and J. Currivan, 2008 – *Cosmos: A Co-Creator's Guide to the Whole World*. Hay House, Inc., Carlsbad, CA.

Leadbeater, Charles W. and A. W. Besant, 2011 - *Occult Chemistry Illustrated Edition: Clairvoyant Observations on the Chemical Elements*. www.zuubooks.com

Lee, Richard H., 1997 – *Bioelectric Vitality: Exploring the Science of Human Energy*. China Healthways Institute, San Clemente, CA.

Linux Today, 2013 – "New 2D Metal Printer is Open Source and Affordable." *Linux Today*, December, 4, 2013. www.linuxtoday.com/infrastructure/new-3d-metal-printer-is-open-source-and-affordable.html

Lipnack, Jessica and J. Stamps, 1986 – *The Networking Book: People Connecting with People*. Routledge & Keegan Paul, NY.

Lipnack, Jessica and J. Stamps, 1997 – *Virtual Teams: Reaching Across Space, Time, and Organizations with Technology*. John Wyley & Sons, Inc., NY.

Llanos, Miguel, 2012 –"Great Barrier Reef Coral Seeing 'Major Decline' Scientists Report." NBC News, October 17, 2012. www.worldnews.nbcnews.com/news/2012/10/01/14152900-great-barrier-reef-coral-seeing-major-decline-scientists-report?lite

Loewen, James W., 1995 – *Lies My Teacher Told Me: Everything Your American History Textbook Got Wrong*. Simon & Schuster, NY.

Lorie, Peter and S. Murray-Clark, 1989 – *History of the Future: A Chronology*. Doubleday, NY.

Louv, Richard, 2012 – *The Nature Principle: Reconnecting with Life in a Virtual Age*. Algonquin Books of Chapel Hill, Chapel Hill, NC.

Lovelock, James E., 1979 – *Gaia: A New Look at Life on Earth*. Oxford University Press, NY.

Lovelock, James E., 1991 – *Healing Gaia: Practical Medicine for the Planet*. Harmony Books, NY.

Loye, David, 1983 – *The Sphinx and the Rainbow: Brain, Mind, and Future Vision*. Shambhala, Boulder, CO.

Lumari, 2003 – *Akashic Records: Collective Keepers of Divine Expression*. Amethyst, Santa Fe, NM.

Machan, Dyan, 2012 – "Bio-Plastic Firms Are Mushrooming." *Smart Ideas*, August 3, 2012, www.smartmoney.com/small-business/small-business/bioplastic-firms-are-mushrooming-1341953846675

Maisel, Eric, 1995 – *A Step-By-Step Guide to Starting and Completing Your Own Work of Art*. Jeremy P. Tarcher/Putnam, NY.

Mann, Charles C., 2002 – "The Test Tube Forest." *Business 2.0,* February, 2002; www.business2.com

Marshall, Michael, 2012 – "Geoengineering with Iron Might Work After All." *New Scientist,* July 21, 2012. www.newscientist.com/article/mg21528744.100-geoengineering-with-iron-might-work-afterall.html#.Uvu60SrDuJk

Martes, C. J., 2006 – *Akashic Field Affirmations: Heal Your Past & Create Your Future*. Martes Group, Inc., Lee's Summit, MO.

Martin, James, 1996 – *Cybercorp: The New Business Revolution*. American Management Association, NY.

Martin, Joel and P. Romanowski, 1997 – *Love Beyond Life: The Healing Power of After-Death Communications*. HarperCollins Publishers, NY.

Maynard, Elliott, 2002 – *Life Management 3000: Success and Survival in the Third Millennium*. Arcos Cielos Research Center, P. O. Box 20069, Sedona, AZ 86341, www.arcoscielos.com

Maynard, Elliott, 2009 – *Transforming the Global Biosphere: Twelve Futuristic Strategies*. Arcos Cielos Research Center, P. O. Box 20069, Sedona, AZ 86341, www.arcoscielos.com

Maynard, Elliott, 2010 – *Future-Science Art: A Unique Paradigm for Creating "Living Artworks."* Arcos Cielos Research Center, Sedona, AZ.

McCullough, Donald, 1998 – *Say Please, Say Thank You: The Respect We Owe One Another*. Perigee Books, NY.

McLuhan, Marshall and Q. Fiore, 1967 – *The Medium is the Massage*. Bantam Books, NY.

McLuhan, Marshall and B. R. Powers, 1989 – *The Global Village: Transformations in World Life and Media in the 21st Century*. Oxford University Press, NY.

McMann, Jean, 1998 – *Altars and Icons: Sacred Spaces in Everyday Life*. Chronicle Books, San Francisco, CA.

McMoneagle, Joseph, 1993 – *Mind Trek: Exploring Consciousness, Time, and Space Through Remote Viewing*. Hampton Roads Publishing Co., Inc., Charlottesville, VA.

McMoneagle, Joseph, 1998 – *The Ultimate Time Machine: A Remote Viewer's Perception of Time, and Predictions for the New Millennium*. Hampton Roads Publishing Company, Inc., Charlottesville, VA.

McMoneagle, Joseph, 2000 – *Remote Viewing Secrets: A Handbook*. Hampton Roads Publishing Company, Inc. Charlottesville, VA.

McNeill, J. R., 2000 – *Something New Under the Sun: An Environmental History of the Twentieth-Century World*. W. W. Norton & Company, NY.

McNiff, Shawn, 1998 – *Trust the Process: An Artist's Guide to Letting Go*. Chronicle Books, San Francisco, CA.

McTaggart, Lynn, 2008 – *The Field: The Quest for the Secret Force of the Universe*. HarperCollins Publishers, NY.

McTaggart, Lynn, 2008 – *The Intention Experiment: Using Your Thoughts to Change Your Life and the World*. Free Press, NY.

McWaters, Barry, 1982 – *Conscious Evolution: Personal and Planetary Transformation*. Evolutionary Press, San Francisco, CA.

Milanovich, Norma J., 1990 – *We The Arcturians*. Athene Publishing, Albuquerque, NM.

Miles, Robert, 2007 –*Fastwalkers; They Are Here: UFO and Alien Disclosure – The Government Cover-Up Finally Exposed*. (DVD). www.fastwalkers.com

Millay, Jean, 1999 – *Multidimensional Mind: Remote Viewing in Hyperspace*. North Atlantic Books, Berkeley, CA.

Miller, David K., 1998 – *Connecting with the Arcturians*. Planetary Heart Publications, Pine, AZ.

Mitchell, Edgar D., 2008 - *The Way of the Explorer: An Apollo Astronaut's Journey Through the Material and Mystical Worlds*. Career Press, Inc., Franklin Lakes, NJ.

Mitchell, John, 1969 – *View Over Atlantis*. Ballantine Books, NY.

Moore, Waveney Ann, 2004 – "Astronaut: We've Had Visitors: The Sixth Man to Walk on the Moon Shares His Unconventional Views." www.sptimes.com/2004/02/18/Neighborhoodtimes/Astronaut__We_ve_had_.shtml

Morgan, James, 2013 – "Amaze Project Aims to Take 3D Printing into 'Metal Age.'" BBC News, October 15, 2013. www.bbc.co.uk/news/science-environment-24528306

Moscowitz, Clara, 2011 – "Extreme Life on Earth: 8 Bizarre Creatures," www.livescience.com/13377-extremophiles-world-weirdest-life.html

Mulvaney, Kieran, 2010 – "Save the Whales; Save the Poop." Yahoo News, October 14, 2010, www.news.discovery.com/earth/save-the-whales-save-the-poop.htm

Murphy, Brendan D., 2012 –" Micro-Psi and String Theory: How Occultists Beat Physicists to the Punch." SERI-Worldwide, www.seri-worldwide.org/id794.html

Murphy, Michael J. and B. Kolsky, 2012. (DVD). *Why in the World Are They Spraying? An Investigation into One of the Many Agendas Associated with Chemtrail/ Geoengineering Programs*, www.whyintheworldaretheyspraying.com

Murray, Peter, 2012 – "Thorium Reactors Being Tested in Norway," www.youtube.com/watch?v=FN0_WKa67uc

Musicforyourmind.com, 2912 – "What is Schumann Resonance?" www.musicyourmind.com/what-is-schumann-resonance

Myers, Norman, Ed., 1984 – *Gaia: An Atlas of Planet Management*. Anchor Press/ Doubleday & Company Inc., Garden City, NY.

Nadler, Gerald and S. Hibino, 1994 – *Breakthrough Thinking: The Seven Principles of Creative Problem Solving*. Prima Publishing, Rocklin, CA.

Nanyang Technical University, 2013 – "NTU Scientist Develops a Multi-Purpose Wonder Material to Tackle Environmental Challenges." March 20, 2013, www.media.ntu.edu.sg/NewsReleases/Pages/newsdetail.aspx?news=14e3b618-c71c 4f20-935c-2a566af5a298

Naisbitt, John, 1994 – *Global Paradox: The Bigger the World Economy, the More Powerful It's Smallest Players*. William Morrow and Company, NY.

Naisbitt, John and P. Aburdene, 1990 – *Megatrends 2000: Ten New Directions for the 1990's*. Wm. Morrow & Co., Inc., NY.

Nani, Christel, 2004 – *Diary of a Medical Intuitive*. Queen's Court Press, Palm Springs, CA.

Nichols, L. Joseph, 2000 – *The Soul as Healer: Lessons in Affirmation, Visualization, and Inner Power*. Llewellyn Publications, St Paul, MN.

Obrien, Terrence, 2010 – "Company Wants to Turn Carbon Pollution Into Baking Soda." Switched, April 25, 2010, www.switched.com/2010/04/25/company-wants-to-turn-carbon-pollution-into-baking-soda

Ohara, Mary and R. Johansen, 1994 – *Global Work: Bridging Distance, Culture & Time*. Jossey-Bass Publishers, San Francisco, CA.

O'Leary, Brian, 2008 – *The Energy Solution Revolution*. Bridger House Publishers, Inc., Hayden, ID.

Onion, Amanda, 2013 – "Project Grows Corals to Repair Reef." ABC News, Dec. 8. 2013, www.abcnews.go.com/Technology/story?id=99439&page=2#.UaT7RyrDuuo

Ontario Consultants on Religious Tolerance, 2010 –"Roman Catholics Beliefs and Practices: Demonic Possession & Oppression; Exorcism," www.religioustolerance.org/chr_exor2.htm

Ornstein, Robert, and P. Ehrlich, 1989 – *New World New Mind: Moving Toward Conscious Evolution*. Doubleday, NY.

Orr, Leonard, 1998 – *Breaking the Death Habit: The Science of Everlasting Life*. Frog, Limited, Berkeley, CA.

Oskin, Becky, 2013 – "What Lives in Antarctica's Buried Lake?" Livescience, December 11, 2013. www.livescience.com/41854-antarctica-lake-whillans-life-results.html

Ostrander, Shelia and L. Schroeder, 1979 – *Super-Learning*. Dell Publishing Company, Inc., NY.

Otterwalks, 2013 – "HAARP: An Overview of Regional Climate Events: Natural Cloud Bows." Planet In fowars, March 25, 2013, www.planet.infowars.com/health/haarp-an-overview-of-regional-climate-events

Ozaniec, Naomi, 1994 – *Dowsing for Beginners*. Hodder and Stoughton, London, UK.

Palloff, Rena M., 1999 – *Building Learning Communities in Cyberspace: Effective Strategies for the Online Classroom*. Jossey-Bass, Inc., Publishers, San Francisco, CA.

Paradigm Research Group, 2013 – www.paradigmresearchgroup.org

Payne, Buryl, 1997 – *Magnetic Healing: Advanced Techniques for the Application of Magnetic Forces*. Lotus Press, Twin Lakes, WI.

Peirce, Penny, 2009 – *Frequency: The Power of Personal Vibration*. Atria Books, Hillsboro, OR.

Penrose, Roger, 1989 – *The Emperor's New Mind: Concerning Computers, Minds, and the Laws of Physics*. Oxford University Press, NY.

Perlman, Lewis J., 1992 – *Schools Out: A Radical New formula for the Revitalization of America's Educational System*. Avon Books, NY.

Perrone, Matthew, 2012 – "Fast-Growing Fish May Never Wind Up on Your Plate." Associated Press, 12/5/12.

Peter, Lawrence J., 1985 – *Why Things Go Wrong or the Peter Principle Revisited*. Bantam Books, NY.

Peter, Lawrence J. and R. Hull, 1969 – *The Peter Principle: Why Things Always Go Wrong*. Bantam Books, NY.

Peters, Tom, 1994 – *The Pursuit of Wow! Every Person's Guide to Topsy Turvy Times*. Random House, NY.

Peters, Tom and N. Austin, 1985 – *A Passion for Excellence: The Leadership Difference*. Warner Books, NY.

Peterson, John L., 1994 – *The Road to 2005: Profiles of the Future*. Waite Group Press, Courte Madera, CA.

Pettis, Chuck, 1999 – *Secrets of Sacred Space: Discover and Create Places of Power*. Llewellyn Press, St. Paul, MN.

Pfeiffer, Eric, 2013 – Yahoo News, *The Sideshow*, "Father of Humankind is 340,000 Years Old." www.news.yahoo.com/blogs/sideshow/father-humankind-340-000-years-old-210033011.html

Phillips, Stephen M., 1980 – *Extra-Sensory Perception of Quarks*. Theosophical Publishing House, Wheaton, IL.

Phys.org News, 2010 – "The Healing Effects of Forests." www.phys.org/news199121152.html

Pickover, Clifford A. - *Computers and the Imagination: Visual Adventures Beyond the Edge*. St. Martin's Press, NY.

Pincott, Karla, 2014 – "Car that Needs Refueling Only Every 100 years." Carsguide.com, January 30, 2014. www.carsguide.com.au/news-and-reviews/car-news/car_that_needs_refueling_only_every_100_years

Pimsleur, Paul, 2012 – www.PimsleurApproach.com www.youtube.com/watch?v=1aYYH56EhYA

Playfair, Guy L. and S. Hill, 1978 – *The Cycles of Heaven: Cosmic Forces and What They Are Doing to You*. Avon Books, NY.

Pogacnik, Marko, 1995 – *Nature Spirits and Elemental Beings: Working with the Intelligence in Nature*. Findhorn Press, Forres, Scotland.

Pollack, Andrew, 2012 – "Engineered Fish Moves a Step Closer to Approval." *New York Times*, 12/21/12.

Poppe, Elisabeth, 2014 – *Resonance – Beings of Frequency*. www.youtube.com/watch?v=-6n-fIHGia8

Potter, Ned, 2012 – "Flowering Plant Revived after 30,000 Years in Russian Permafrost." ABC News, Feb. 20, 2012, www.abcnews.go.com/blogs/technology/2012/02/flowering-plant-revived-after-30000-years-in-russian-permafrost

Powell, Tag and C. H. Mills, 1993 – *ESP for Kids: How to Develop Your Child's Psychic Ability*. Top of the Mountain Publishing, Key Largo, FL.

Pritchard, Ray, 1998 – *Something New Under the Sun: Ancient Wisdom for Contemporary Living*. Moody Press, Chicago, IL.

Psi Tech, 2014 – "The World's Most Powerful Information Collection & Intelligence Technology," www.psitech.net

Rachele, Sal, 2008 – *Earth Changes and 2012: Messages from the Founders*. Living Awareness Productions, Sedona, AZ.

Rachele, Sal, 2013 – Professional Channeler. www.salrachele.com

Radin, Dean, 1997 – *The Conscious Universe: The Scientific Truth of Psychic Phenomena*. Harper-Collins Publishers, NY.

Randour, Mary Lou, 2000 – *Animal Grace: Entering Into a Spiritual Relationship with Our Fellow Creatures*. New World Library, Novato, CA.

Radin, Dean, 1997 – *The Conscious Universe: The Scientific Truth of Psychic Phenomenon*. HarperCollins Publishers, NY.

Radin, Dean, 2006 – *Entangled Minds: Extrasensory Experiences in a Quantum Reality*. Paraview Pocket Books, NY.

Raffill, Thomas E., 2000 – "China's SuperPsychics Revisited." *Spirit of Ma'at,* Vol. 1, October, 2000.

Reed, Jack, 2001- *The Next Evolution: A Blueprint for Transforming the Planet*. The Community Planet Foundation, Santa Barbara, CA.

Rettner, Rachel, 2008 – "Amazing Power of Music Revealed." www.livescience.com/health/081015-music-power.html

Rettner, Rachael, 2010 – "Music 'Tones the Brain, Improves Learning." www.news.yahoo.com/s/livescience/20100720/sc_livescience/musictonesthebrain improveslearning

Rheingold, Howard, 1991 – *Virtual Reality: The Revolutionary Technology of Computer-Generated Artificial Worlds, and How it Promises and Threatens to Transform Business and Society*. Summit Books, NY.

Richards, Dick, 1998 – *Setting Your Genius Free: How to Discover Your Spirit and Calling*. Berkley Books, NY.

Roads, Michael J., 1985 – *Talking With Nature: Sharing the Energies and Spirit of Trees, Plants, Birds, and Earth*. H. J. Kramer, Inc., Tiburon, CA.

Robbins, Dianne, 1997 –*The Call Goes Out: Messages from Earth's Cetaceans*. Inner Eye Books, Livermore, CA.

Robbins, John, 1987 – *Diet for a New America: How Your Food Choices Affect Your Health, Happiness and the Future of Life on Eart*h. H. J. Kramer, Tiburon, CA.

Rockoff, Jonathan D., 2013 – "Life Discovered in the Deepest Ocean." *The Wall Street Journal*, Monday, March 18, 2013.

Roman, Joseph, A. Rosenfeld, and S. Herman, 1999. *The Internet Economy and Global Warming*. The Center for Energy and Climate. www.cool-companies.org

Rosenthal, Allen M., 1991 – *Your Mind the Magician*. DeVorss & Company, Marina del Ray, CA.

Royal, Lissa, 1997 – *Millennium: Tools for the Coming Changes*. Royal Priest Research Press, Phoenix, AZ.

Russell, Peter, 1983 – *The Global Brain: Speculations on the Evolutionary Leap to Planetary Consciousness*. J.P. Tarcher, Inc., Los Angeles, CA.

Russell, Peter, 1995 – *The Global Brain Awakens: Our Next Evolutionary Leap*. Global Brain, Inc., Palo Alto, CA.

Sagan, Carl, 1977 – *The Dragons of Eden: Speculations on the Evolution of Human Intelligence*. Ballantine Books, NY.

Salim, Emil and O. Ullsten, Eds, 1999 – *Our Forests, Our Future: Report of the World Commission on Forests and Sustainable Development*. Cambridge University Press, NY.

Sanaya, Roman and D. Packer, 1987 – *Opening to Channel: How to Connect With Your Spirit Guide*. H. J. Kramer, Inc., Tiburon, CA.

Sanger, Johanna, 2012 – "Musical Duets Lock Brains as well as Rhythms." *Frontiers of Neuroscience*, Nov. 29, 2012, www.eurekalert.org/pub_releases/2012-11/f-mdl112712.php

Saraydarian, Torkom, 1966 – *The Creative Fire*. T. S. G. Publishing Foundation, Inc. Cave Creek, AZ.

SCENAR-Cosmodic, 2014 – "L E T Medical VX 735AG." www.scenar.us/ic/080002.html

Schneider, Stephen H., 1997 – *Laboratory Earth: The Planetary Gamble We Can't Afford to Lose*. Basic Books, NY.

Schonwald, Josh, 2012 – *The Taste of Tomorrow: Dispatches from the Future of Food*. HarperCollins Publishers, NY.

Schwartz, Ariel, 2010 – "Urbee Hybrid: The First 3D Printed Car." *Fast Company*, Nov. 1, 2010, www.fastcompany.com/1698943/urbee-hybrid-first-3-d-printed-car

Schwartz, Gary E., W. L Simon and L. G. S. Russek, 1999 – *The Living Energy Universe: A Fundamental Discovery That Transforms Science and Medicine*. Hampton Roads Publishing Company, Inc., Charlottesville, VA.

Schwartz, Gary E., W. L Simon and L. G. Russek, 2001 – *The Afterlife Experiments: Breakthrough Scientific Evidence of Life after Death*.

Schwartz, Joel, 2003 – "World Gets Bleaker for our Children, Contends Psychologist." January, 27, 2003; http://www.washington.edu/newsroom.

Schwartz, Stephan. A., 1978 – *The Secret Vaults of Time: Psychic Archaeology and the Quest for Man's Beginnings*. Hampton Roads Publishing Company, Charlottesville, VA.

Science Daily, 2008 – "World's Oldest Living Tree – 9550 Years Old – Discovered in Sweden." *Science Daily*, March 16, 2013.

Scott-Mumby, 2014 – "Star Trek Medicine Is Here! Now!!" February 27, 2014. www.alternative-doctor.com/specials/scenar.htm

Seeger, Hilbert, 2007 – (Video): "Live Blood Analysis Dark Field Microscopy Explained." October 26, 2007, www.youtube.com/watch?v=Mg21897IW_4

Seeger, Hilbert, 2014 – (Video): "Live Blood Cell Analysis with a Dark Field Electron Microscope." October 9, 2013. www.youtube.com/watch?v=Iu2_kZOg6ak

Sereda, David and Jim Law, 2009 – *Quantum Communication: Unleash the Powers of Your Mind*. (DVD). Voice Entertainment, Sedona, AZ.

Sergo, Peter, 2007 – "Greening Up the Ocean: Environmental Business Sets Sail with Hopes of Creating a New Frontier of Ocean Carbon Storage." *Scienceline*, June 8, 2007, www.scienceline.org/2007/06/environment-sergo-carbonsequestration

Shane, Harold G., 1987 – *Teaching and Learning in a Microelectronic Age*. Phi Delta Kappa Educational Foundation, Bloomington, IN.

Shapiro, Andrew L, 1999 – *The Control Revolution: How the Internet in Putting Individuals in Charge and Changing the World We Know*. Century Foundation Books, NY.

Sheldrake, Rupert, 1988 – *The Presence of the Past: Morphic Resonance and the Habits of Nature*. Vintage Books, NY.

Sheldrake, Rupert, 1988 – *The Rebirth of Nature: The Greening of Science and God*. Bantam Books, NY.

Shepherd, Claire, 2011 – Video: "Live Blood Analysis." August 24, 2011. www.youtube.com/watch?v=kCc18eWs5XU

Shuman, Bruce A., 1989 – *The Library of the Future*. Libraries Unlimited, Inc., Englewood, CO.

Simpson, Liz, 2000 – *The Healing Energies of Earth*. Journey Editions, Boston, MA.

Simpson, Rick, 2009 – (Video) *Run from the Cure: The Rick Simpson Story*. www.youtube.com/watch?v=0psJhQHkGI

Simpson, Rick, 2012 – (Video), *Canada's New Oil Men: A Phoenix Tears Story*. www.phoenixtears.ca

Singer, Emily, 2007 – "Stem Cells Without the Embryos: An Easy Method for Reprogramming Adult Cells May Resolve Ethical Objections." *MIT Technology Review*, Nov. 20, 2007, www.technologyreview.com/news/409077/stem-cells-without-the-embryos

Slade, Dea and M. Radman, 2011 – "Oxidative Stress Resistance in *Deinococcus radiodurans*." *Microbiology and Molecular Biology Reviews*, March, 2011, v. 75(1).

Slouka, Mark, 1995 – *War of the Worlds: Cyberspace and the High-Tech Assault on Reality*. Basic Books, NY.

Small, Jacquelyn, 1982 – *Transformers: The Artists of Self-Creation*. DeVorss & Company, Marina del Rey, CA.

Small, Jacquelyn, 1995 – *Becoming a Practical Mystic: Creating Purpose for Our Spiritual Future*. Theosophical Publishing House, Wheaton, IL.

Smil, Vaclav, 1999 – *Energies: An Illustrated Guide to the Biosphere and Civilization*. MIT Press, Cambridge, MA.

Smith, Penelope, 1999 – *When Animals Speak: Animal Interspecies Telepathic Communication*. Beyond Words Publishing, Inc., Hillsboro, OR.

Solomon, Daniel L., Rauland-Borg Corporation, 2004 – "The Difference Between Dirac's Hole Theory and Quantum Field Theory." *Frontiers in Quantum Physics Research*. Nova Science, Hauppauge, NY.

Sonora, Terra, 2013 – Professional Channeler. www.angelchannel.com

Stableford, Brian, 1984 – *Future Man: Brave New World, or Genetic Nightmare*. Crown Publishers, Inc., NY.

Starr, Mark, 2014 – "Magnetico Sleep Systems." www.type2hypothyroidism.com/magnetico.html

Stearn, Jess, 1976 – *The Power of Alpha Thinking - A Miracle of the Mind*. New American Library, NY.

Stefik, Mark, 1986 – *Internet Dreams: Archetypes, Myths, and Metaphors*. MIT Press, Cambridge, MA.

Stefik, Mark, 1999 – *The Internet Edge: Social, Legal, and Technological Challenges for a Networked World*. MIT Press, Cambridge, MA.

Steinberg, Laurence, 1996 – *Beyond the Classroom: Why School Reform Has Failed, and What Parents Need to Do*. Simon & Schuster, NY.

Stock, Gregory, 1993 – *Metaman: The Merging of Humans and Machines into a Global Superorganism*. Simon & Schuster, NY.

Sussman, Janet I., 1996 – *Timeshift: The Experience of Dimensional Change*. Timeportal Publications, Fairfield, IA.

Tansley, David D., 1977 – *Dimensions of Radionics: Techniques of Instrumented Distance-Healing, a Manual of Radionic Theory and Practice*. Brotherhood of Life, Inc., Albuquerque, NM.

Targ, Russell, and J. Katra, 1998 – *Miracles of the Mind: Exploring Nonlocal Consciousness and Spiritual Healing*. New World Library, Novato, CA.

Taylor, David C., 2014 – "Man-Made Earthquake and Tsunami Warning Issued for California and Hawaii." www.truther.org, February 10, 2014, www.truthernews.wordpress.com/2014/02/10/man-made-earthquake-and-tsunami-warning-issued-for-california-hawaii

Telesco, Trish, 1998 – *Dancing With Devas: Connecting with the Spirits and Elements of Nature*. Belfrey Books, Laceyville, PA.

Thornburg, David D., 1993 – *Education in the Communications Age*. Thornburg Center, San Carlos, CA.

Thurnell-Read, Jane, 1995 – *Geopathic Stress: How Earth Energies Affect Our Lives*. Element Books, Ltd., Boston, MA.

Tiffen, John and L. Rajasingham, 1995 – *In Search of the Virtual Class: Information in an Information Society*. Routledge, NY.

Toffler, Alvin, 1970 – *Future Shock*. Random House, NY.

Toffler, Alvin, 1984 – *Previews and Premises*. South End Press, Boston, MA.

Toffler, Alvin, 1990 – *Power Shift: Knowledge, Wealth, and Violence at the Edge of the 21st Century*. Bantam Books, NY.

Toffler, Alvin and H. Toffler, 1994 – *Creating a New Civilization: The Politics of the Third Wave*. Turner Publishing, Inc., Atlanta, GA.

Trivedi Foundation, 2010 – www.trivedifoundation.org

Tompkins, Peter, 1997 – *The Secret Life of Nature: Living in Harmony with the Hidden World of Nature Spirits from Fairies to Quarks*. HarperCollins Publishers, NY.

Tyberonn, 2007 – *Earth-Keeper: The Energy and Geometry of Sacred Sites; Grids, Vortexes, and Portals of Gaia, the Living Earth.* Star Quest Publishing, Reno, NV.

Turner, John L., 2009 – *Medicine, Miracles, & Manifestations: A Doctor's Journey Through the Worlds of Divine Intervention, Near-Death Experiences, and Universal Energy.* The Career Press, Inc., Franklin Lakes, NJ.

Vickers, Ed, 2010 – "Music and Consciousness." www.sfxmachine.com/docs/musicandconsciousness.html

Vitale, Barbara M., 1982 – *Unicorns are Real: A Right-Brained Approach to Learning.* Jalimar Press, Rolling Hills Estates, CA.

Vitale, Barbara M., 1986 – *Free Flight: Celebrating Your Right Brain.* Jalimar Press, Rolling Hills Estates, CA.

Walters, J. Donald, 1986 – *Education for Life.* Ananda Publications, Nevada City, CA.

Wasserman, Harvey, 1983 – *America Born and Reborn.* Collier Books, NY.

Watson, Lyall, 1973 – *Supernature: A Natural History of the Supernatural.* Hodder and Stoughton, London, UK.

Watson, Lyall, 1979 – *Lifetide: A Biology of the Unconscious.* Hodder and Stoughton, London, UK.

Watson, Lyall, 1988 – *Beyond Supernature: A New Natural History of the Supernatural.* Hodder and Stoughton, London, UK.

Webre, Alfred L., 2005 – *Exopolitics: Politics, Government, and Law in the Universe.* Universe Books, Vancouver, B.C., Canada.

Webre, Alfred L., 2013 – "Prima Facie Forensic Evidence of Weather Warfare Attack in June, 2013 Calgary Floods: Canadian Media and Officials Silent." EcologyNews.com, June 26, 2013. www.exopolitics.blogs.com/peaceinspace/2013/06/by-alfred-lambremont-webre-jd-med-vancouver-bc-there-is-now-a-sufficient-threshold-of-prima-facie-evidence-to-reasonabl.html

Weenolsen, Patricia, 1996 – *The Art of Dying: The Only Book for Persons Facing Their Own Death.* St. Martin's Griffin, NY.

Weintraub, Sandra, 1998 – *The Hidden Intelligence: Innovation Through Intuition.* Butterworth-Heinemann, Woburn, MA.

Welch, Chris, 2013 – "3D Printed Metal Gun Successfully Fires Over 50 Rounds." The Verge, November 7, 2013. www.theverge.com/2013/11/7/5077718/worlds-first-3d-printed-metal-gun-fires-over-50-rounds

Wheatley, Margaret, 1992 – *Leadership and the New Science: Learning about Organization from an Orderly Universe.* Berrett-Koehler Publishers, Inc., San Diego, CA.

White, Frank, 1987 – *The Overview Effect.* Houghton Mifflin Company, Boston, MA.

White, Ken W. and B. H. Weight, 2000 – *The Online Teaching Guide: A Handbook of Attitudes, Strategies, and Techniques for the Virtual Classroom.* Allyn and Bacon, Boston, MA.

White, John and S. Krippner, 1977 – *Future-Science: Life Energies and the Physics of Paranormal Phenomena*. Doubleday and Company, Inc., NY.

Whitford, Gwenith, 2000 – "Coral Farming: Striving to Revive the 'Underwater Rainforests." *Caribbean Compass*, December, 2000. www.caribbeancompass.com/coral.htm

Whittle, David B., 1997 – *Cyberspace: The Human Dimension*. W. H. Freeman and Company, NY.

Wikipedia, 2012 – "3D Printing." www.en.wikipedia.org/wiki/3D_printing

Wikipedia, 2013, Methuselah (tree) - www.en.wikipedia.org/wiki/Methuselah(tree)

Wikipedia, 2013, Edgar Mitchell - www.en.wikipedia.org/wiki/Edgar_Mitchell#Views_on_UFOs

Wilde, Steward, 2000 – *Sixth Sense: Including the Secrets of the Etheric Subtle Body*. Hay House, Inc., Carlsbad, CA.

Williams, Linda V., 1983 – *Teaching for the Two-Sided Mind: A Guide to Right Brain/Left Brain Education*. Simon & Schuster, NY.

Winchester, Edward, 2013 – Founder of the Pentagon Meditation Club. www.pentagonmeditationclub.com/history_about_us.htm, personal communications.

Winner, Langdon, 1986 – *The Whale and the Reactor: A Search for Limits in an Age of Technology*. W. H. Freeman and Company, NY.

Wittels, Nedda, 2010 – "What Animal Communicators Do." About.com, Veterinary Medicine: The Viewer Viewpoint, www.vetmedicine.about.com/od/behaviortraining/a/Wittels072304.htm

Wonder, Jacquelyn and P. Donovan, 1984 – *Whole-Brain Thinking: Working from Both Sides of the Brain to Achieve Peak Job Performance*. Ballentine Books, NY.

Wood, George H., 1992 – *Schools That Work: America's Most Innovative Public Education Systems*. Penguin Books, NY.

Wood, Robert M. and N. Redfern, 2013 – *Alien Viruses: Crashed UFOs, MJ-12 & Biowarfare*. Richard Dolan Press, Rochester, NY.

Woolf, V. Vernon, 1990 – *Holodynamics: How to Develop and Manage Your Personal Power*. Self-Published by V. Vernon Woolf.

Woolf, V. Vernon, 2004 – *The Dance of Life: Transform Your World NOW! Create Wellness, Resolve Conflicts, and Learn to Harmonize Your "Being" with Nature*. The International Academy of Holodynamics, Reno, NV.

World Wildlife Fund, 2013 – "Valuing Wildlife: One of the Most Critical Issues of our Time." www.nxtbook.com/nxtbooks/wwf/focus_050612/index.php?startid=4

Wyke, Alexandra, 1997 – *21st Century Miracle Medicine: RoboSurgery, Wonder Cures, and the Quest for Immortality*. Plenum Trade, NY.

Yankelovich, Daniel, 1981 – *New Rules: Searching for Self-Fulfillment in a World Turned Upside Down*. Bantam Books, NY.

Yatri, 1988 – *Unknown Man: The Mysterious Birth of a New Species.* Simon & Schuster, Inc., NY.

Yoder, Susan and M. Benton, 2000 – *The Wisdom of Dolphins.* Sourcebooks, Inc., Naperville, IL.

Youngblood, Gene, 1970 – *Expanded Cinema.* E. P. Dutton and Co., NY.

YouTube, 2010 – *HAARP Holes in Heaven* (Video). December 27, 2010. www.youtube.com/watch?v=10DihgRIs0I&list=PL99AB5E3E70406BF0

Zey, Michael, 1998 – *Seizing the Future: How the Coming Revolution in Science, Technology, and Industry Will Expand the Frontiers of Human Potential and Reshape the Planet.* Simon & Schuster, NY.

Zukov, Gary, 1979 – *The Dancing Wu Li Masters: An Overview of the New Physics.* Wm. Morrow and Company, Inc., NY.